Machining and machine-tools

WOODHEAD PUBLISHING REVIEWS:

MECHANICAL ENGINEERING

Series Editor: Professor J. Paulo Davim,
Department of Mechanical Engineering, University of Aveiro,
Portugal and Head of MACTRIB – Machining and Tribology Research Group
(email: pdavim@ua.pt)

Woodhead Publishing is pleased to publish this major Series of books entitled *Woodhead Publishing Reviews: Mechanical Engineering*. The Series Editor is Professor J. Paulo Davim, Department of Mechanical Engineering, University of Aveiro, Portugal and Head of MACTRIB – Machining and Tribology Research Group.

This research Series publishes refereed, high-quality articles with a special emphasis on research and development in mechanical engineering from a number of perspectives including (but not limited to): machining and machine tools; tribology and surface engineering; materials and manufacturing processes; solid mechanics and structural mechanics; computational mechanics and optimization; mechatronics and robotics; fluid mechanics and heat transfer; renewable energies; biomechanics; micro- and nano-mechanics, etc. We seek authors, editors and contributors from a broad range of areas within the mechanical engineering discipline.

This Series examines current practices and possible future developments within the research field and industry-at-large. It is aimed at an international market of academics, practitioners and professionals working in the area. The books have been specially commissioned from leading authors, with the objective of providing the reader with an authoritative view of current thinking.

New authors: we would be delighted to hear from you if you have an idea for a book. We are interested in both shorter, practically orientated publications (45 000+ words) and longer, theoretical monographs (75 000–100 000 words). Our books can be single, joint or multi-author volumes. If you have an idea for a book, please contact the publishers or Professor J. Paulo Davim, the Series Editor.

Published by Woodhead Publishing Limited, 2013

Dr Glyn Jones
Woodhead Publishing Limited
Email: gjones@chandospublishing.com
www.woodheadpublishing.com

Professor J. Paulo Davim
Department of Mechanical Engineering,
University of Aveiro, Portugal
Email: pdavim@ua.pt
http://www2.mec.ua.pt/machining/pers-davim.htm

Woodhead Publishing Limited: established in 1989, Woodhead Publishing is a leading independent international publisher, publishing in the following main areas: food science, technology and nutrition; materials engineering; welding and metallurgy; textile technology; environmental technology; finance, commodities and investment; and mathematics.

Our ambitious publishing plans for the future will continue to bring you a range of authoritative reference books, professional texts and monographs, all written and produced to the exacting standards that have made Woodhead Publishing one of the UK's fastest growing independent publishers. All of our books are written in direct response to customers' needs by a truly international team of authors, ensuring they are designed for and relevant to a global audience.

Woodhead Publishing books are available worldwide, either direct, via our website, www.woodheadpublishing.com, or through booksellers and an international network of agents and representatives.

Professor J. Paulo Davim received his PhD in Mechanical Engineering from the University of Porto in 1997 and the Aggregation from the University of Coimbra in 2005. Currently, he is Aggregate Professor in the Department of Mechanical Engineering of the University of Aveiro and Head of MACTRIB – Machining and Tribology Research Group. He has more than 25 years of teaching and research experience in manufacturing, materials and mechanical engineering with special emphasis in machining and tribology. He is the Editor of five international journals, and also guest editor, editorial board member, reviewer and scientific advisor for many international journals and conferences. He has also published, as author and co-author, more than 40 book chapters and 350 articles in ISI journals (h-index 25+) and conferences.

Bulk orders: some organizations buy a number of copies of our books. If you are interested in doing this, we would be pleased to discuss a discount. Please email wp@woodheadpublishing.com or telephone +44 (0) 1223 499140.

Published by Woodhead Publishing Limited, 2013

Related titles:

Mechatronics and manufacturing engineering: Research and development
Number 1 in the Woodhead Publishing Reviews: Mechanical Engineering Series
(ISBN 978-0-85709-150-5)
Mechatronics is the blending of mechanical, electronic and computer engineering into an integrated design, and has an increasing impact on engineering processes, particularly on the design, development and operation of manufacturing systems in a wide range of industries. *Mechatronics and manufacturing engineering* is the first in the **Woodhead Publishing Reviews: Mechanical Engineering Series** and presents high quality articles with a special emphasis on research and development in mechatronics and manufacturing engineering. The book is divided into eight chapters, covering implementation of light-scattering instrumentation; planar micromanipulation on microconveyor platforms; basic active and passive joint torque control in a single-axis arm; signal processing for tool condition monitoring; ANN modelling of fractal dimension in machining; predicting forces and damage in drilling of polymer composites; minimising burr size in drilling; and single point incremental forming of polymers.

Materials and surface engineering: Research and development
Number 2 in the Woodhead Publishing Reviews: Mechanical Engineering Series
(ISBN 978-0-85709-151-2)
Materials science is an interdisciplinary field concerned with the micro and nanostructure of materials, and their properties, processing and applications in engineering, technology and industry. *Materials and surface engineering* is the second in the **Woodhead Publishing Reviews: Mechanical Engineering Series**, presenting high quality articles with a special emphasis on research and development in materials and surface engineering and the resultant applications. The book is divided into eight chapters, covering scratch resistance of modified polymethylmethacrylate nanocomposites; nanomechanical properties and nanoscale deformation of engineering materials and alloys; analysis of large-strain microindentation of crystalline polymers; nanocomposite coatings; thermal sprayed Wc-Co coatings; tribological performance of electroless Ni-P coatings, tribological response of materials during sliding; and temperature and stress fields in laser straight cutting of Ti-6Al-4V alloy.

The design and manufacture of medical devices
Number 4 in the Woodhead Publishing Reviews: Mechanical Engineering Series
(ISBN 978-1-907568-72-5)
Medical devices play an important role in the field of medical and health technology, and encompass a wide range of health care products. Directive 2007/47/EC defines a medical device as *any instrument, apparatus, appliance, software, material or other article, whether used alone or in combination, including the software intended by its manufacturer to be used specifically for diagnostic and/or therapeutic purposes and necessary for its proper application, intended by the manufacturer to be used for human beings. The design and manufacture of medical devices*, the fourth in the **Woodhead Reviews: Mechanical Engineering Series**, brings together a range of articles and case studies dealing with medical device R&D. Chapters in the book cover materials used in medical implants, such as Titanium Oxide, polyurethane, and advanced polymers; devices for specific applications such as spinal and craniofacial implants, and other issues related to medical devices, such as precision machining and integrated telemedicine systems.

Details of these and other Woodhead Publishing books can be obtained by:

- visiting our web site at www.woodheadpublishing.com
- contacting Customer Services (e-mail: sales@woodheadpublishing.com; fax: +44(0) 1223 832819; tel: +44(0) 1223 499140; address: Woodhead Publishing Limited, 80 High Street, Sawston, Cambridge CB22 3HJ, UK)

If you would like to receive information on forthcoming titles, please send your address details to Customer Services, at the address above. Please confirm which subject areas you are interested in.

Published by Woodhead Publishing Limited, 2013

Woodhead Publishing Reviews: Mechanical Engineering

Machining and machine-tools

Research and development

EDITED BY
J. PAULO DAVIM

WP
WOODHEAD
PUBLISHING

Oxford Cambridge Philadelphia New Delhi

Published by Woodhead Publishing Limited, 2013

Woodhead Publishing Limited, 80 High Street, Sawston, Cambridge CB22 3HJ, UK
www.woodheadpublishing.com
www.woodheadpublishingonline.com

Woodhead Publishing, 1518 Walnut Street, Suite 1100, Philadelphia, PA 19102-3406, USA

Woodhead Publishing India Private Limited, G-2, Vardaan House, 7/28 Ansari Road, Daryaganj, New Delhi – 110002, India
www.woodheadpublishingindia.com

First published in 2013, Woodhead Publishing Limited

British Library Cataloguing-in-Publication Data
A catalogue record for this book is available from the British Library

Library of Congress Control Number: 2012954745

Woodhead Publishing ISBN: 978-0-85709-154-3 (print)
ISBN: 978-0-85709-219-9 (online)

Woodhead Publishing Reviews: Mechanical Engineering
ISSN: 2048-0571 (print) and ISSN: 2048-058X (online)

Typeset by Domex e-Data Pvt. Ltd., India
Printed in the UK and USA.

Contents

List of figures and tables — *xi*
List of abbreviations — *xxiii*
Preface — *xxv*
About the contributors — *xxvii*

1 Analysis of acoustic emission signal evolutions for monitoring diamond-coated tool delamination wear in machining — **1**
P. Lu and Y. K. Chou
Introduction — 2
Experimental set-up and methods — 6
Results and discussion — 12
Conclusions — 24
Acknowledgments — 26
References — 26

2 High-performance machining of austenitic stainless steels — **29**
Ana Isabel Fernández-Abia, Joaquín Barreiro García and Luis N. López de Lacalle
Introduction — 30
Properties of austenitic stainless steels — 33
Machining study of austenitic stainless steels at high performance conditions — 43
Conclusions — 86
References — 88

Published by Woodhead Publishing Limited, 2013

3	**Forces monitoring in shape grinding of complex parts**	**91**
	A. Di Ilio	
	Introduction	92
	Generality of screw rotors	95
	Workpiece holder	97
	Experimental tests	101
	Results and discussion	104
	Conclusions	109
	Acknowledgements	109
	References	109
4	**Optimization of minimum quantity lubrication in grinding with CBN wheels**	**113**
	E. C. Bianchi, P. R. Aguiar, R. C. Canarim and A. E. Diniz	
	Introduction	113
	Literature review	116
	Wheel cleaning	122
	Conclusions	129
	References	129
5	**Electrical discharge machining: study on machining characteristics of WC/Co composites**	**135**
	K. Palanikumar and J. Paulo Davim	
	Introduction	136
	Principles of the EDM process	140
	Machining experiments	145
	Response surface modelling using Box–Behnken design	149
	Response surface model for MRR and surface roughness	152
	Results and discussion	156
	Summary	164
	Acknowledgements	165
	References	165

Published by Woodhead Publishing Limited, 2013

6 **Conventional and unconventional hole making in metal matrix composites** **169**

Inderdeep Singh, Sarbjit Singh and Abhishek Singh

Introduction 170

Development of the secondary processes for MMCs: 'need of the hour' 170

Classification of MMCs based on the matrix material 171

Fabrication processes for MMCs 172

Conventional and unconventional machining processes 174

Conventional secondary processing of MMCs 175

Hole making in MMCs 176

Unconventional machining of MMCs 182

References 191

7 **A laboratory machine for micro electrochemical machining** **195**

G. R. Ribeiro, I. M. F. Bragança, P. A. R. Rosa and P. A. F. Martins

Introduction 195

Development of the laboratory machine 199

Experimental work plan 202

Results and discussion 204

Conclusions 209

Acknowledgment 209

References 210

8 **Cam-driven electromagnetic mechanical testing machine** **211**

C. M. A Silva, P. A. R. Rosa and P. A. F. Martins

Introduction 212

Design and fabrication of the testing machine 215

Experimental workplan 222

Results and discussion 223

Conclusions 229

Acknowledgment 230

References 230

Index **233**

List of figures and tables

Figures

1.1	Experimental set-up in a CNC lathe showing an AE sensor mounted on the tool holder (taken from Hu *et al.*, 2008)	7
1.2	AE-FFT intensity (high and low frequencies) with cutting time for insert F	9
1.3	AE-FFT at different time periods for insert F during the failure pass: (a) first, (b) second, (c) third and (d) last 25 per cent cutting time of the entire machining pass	10
1.4	AE-FFT amplitude comparisons in the four sub-periods during the coating failure pass of insert F	11
1.5	Illustration of the short-time Fourier transform (STFT) method (from Lu *et al.*, 2010)	12
1.6	Raw AE signal for insert A at: (a) initial cutting pass and (b) tool failure pass	13
1.7	AE-RMS amplitude during initial cutting and tool failure pass for insert A (from Lu *et al.*, 2009)	14
1.8	AE-RMS during tool failure pass: (a) with clear failure transition period (insert A) and (b) without clear failure period (insert B) (from Lu *et al.*, 2009)	15

1.9 FFT spectra of raw AE signals from two cutting passes of insert A: (a) initial cutting and (b) final cutting (from Lu *et al.*, 2009) 15

1.10 AE-FFT intensity of high- and low-frequency components with cutting time for (a) insert A and (b) insert B (from Lu *et al.*, 2009) 16

1.11 AE-FFT of insert A: (a) initial cutting pass and (b) coating failure pass 17

1.12 AE-FFT of insert B: (a) initial cutting pass and (b) coating failure pass (from Lu *et al.*, 2009) 17

1.13 AE-FFT at different time periods for insert A during the coating failure pass: (a) first, (b) second, (c) third and (d) last 25 per cent cutting of the entire pass (from Lu *et al.*, 2009) 18

1.14 AE-FFT amplitude comparisons, between the low- and high-frequency components, in the four sub-periods during the coating failure pass: (a) insert A and (b) insert B (from Lu *et al.*, 2009) 18

1.15 Amplitude ratio of the high/low-frequency components by the STFT method during different cutting passes of insert A: (a) initial cutting passes, (b) prior to failure pass and (c) failure pass (from Lu *et al.*, 2010) 20

1.16 Amplitude ratio of high/low-frequency components by the STFT method during coating failure passes of insert B (from Lu *et al.*, 2010) 20

1.17 Tool wear development of cutting inserts (A, B and C) with cutting time 21

1.18 Amplitude ratios of high/low-frequency components by the STFT method during coating failure pass for insert C (from Lu *et al.*, 2010) 22

1.19 Tool wear development of inserts D, E and F with cutting time (from Lu *et al.*, 2010) 23

1.20 Amplitude ratio of high/low-frequency components during the coating failure pass: (a) insert D, (b) insert E and (c) insert F (from Lu *et al.*, 2010) 24
2.1 World production of stainless steel in the last decade 31
2.2 Effect of alloy elements on AISI304 austenitic stainless steel 35
2.3 Photomicrographs of AISI303 (×200) 45
2.4 Scheme of the experimental equipment 46
2.5 Directions of the three components of cutting force acquired by a Kistler 9121 dynamometer 46
2.6 Areas for roughness measurement in the machined specimen 47
2.7 (a) Equipment for image acquisition, and (b) software developed for tool wear measurement 48
2.8 Orthogonal cutting scheme 50
2.9 Comparison between simulated and experimental tangential forces over a wide range of cutting conditions 53
2.10 (a) Relationship between cutting force and cutting speed, and (b) normalized cutting forces 54
2.11 Relationship between shearing angle and cutting speed 54
2.12 Effect of cutting speed on tool temperature 55
2.13 Thermal energy by unit of volume with respect to V_c 56
2.14 Relationship between temperature in the tool–chip interface and V_c 56
2.15 Plastic strain rate versus cutting speed (V_c) 57
2.16 Chip formation versus cutting speed (f_n = 0.1 mm/rev) 58

Published by Woodhead Publishing Limited, 2013

2.17 Chip thickness versus cutting speed 59

2.18 Effect of non-deformed chip thickness on specific cutting energy (P_s) 61

2.19 Relationship between cutting forces and feed rate 61

2.20 Probe, photomicrograph and micro-hardness measurement for AISI303 63

2.21 Relationship between tangential cutting force and cutting speed (f_n = 0.2 mm/rev; a_p = 1 mm) 64

2.22 Relationship between cutting force amplitude and cutting speed 67

2.23 Scanning electron micrographs of the machined surfaces at different cutting speeds 69

2.24 Relationship between chip thickness and cutting speed 70

2.25 Average microhardness values 71

2.26 Depth of deformed microstructure for specimens machined at different cutting speeds 72

2.27 SEM mapping image of adhered material in the tool flank face 73

2.28 EDX chemical composition analysis for new and worn cutting tools 74

2.29 Scanning electron micrographs showing tool flank wear at several cutting speeds 75–6

2.30 Scanning electron micrographs showing crater wear at several cutting speeds: (a) V_c = 37 m/min; (b) V_c = 450 m/min; (c) V_c = 870 m/min. (d) EDX analysis of areas A and B. 78–9

2.31 Micrograph of a chip for AISI303 stainless steel: (a) morphology of chip under orthogonal cutting with V_c = 60 m/min and f_n = 0.2 mm/rev; (b) portion of chip 81

2.32	Scanning electron micrographs of chips at several cutting speeds	82
2.33	Chip sliding surface at several cutting speeds	83–4
3.1	A pair of asymmetric rotors. (a) Schematic drawing, (b) finished parts. Left, male; right, female	96
3.2	Schematic of the workpiece holder assembly	98
3.3	Calibration scheme for radial forces and torque	99
3.4	Calibration curves for axial force and torque	101
3.5	Force trends as a function of the workpiece rotation angle, when the *x*-axis of the sensor is oriented vertically: (a) experimental, (b) theoretical	101
3.6	Workpiece mounted in a grinding machine	102
3.7	Force components and torque during idle strokes without material removal	104
3.8	Trends of force components and torque as a function of the rotation angle of the workpiece: wheel speed = 32 m/s, feedrate = 1500 m/min, depth of cut = 0.05 mm	106
3.9	Trends of force components and torque as a function of the rotation angle of the workpiece: wheel speed = 32 m/s, feedrate = 3000 m/min, depth of cut = 0.1 mm	106
3.10	Trends of force components and torque as a function of the rotation angle of the workpiece: wheel speed = 38 m/s, feedrate = 3000 m/min, depth of cut = 0.1 mm	107
3.11	Main effects plot of the grinding parameters for torque	108
4.1	Wheel clogging	123
4.2	Grinding wheel with lodged chips (100×)	124

4.3	Effect of compressed air jet on wheel cleaning	127
5.1	Principle of the electric discharge machining process	141
5.2	Schematic of electric discharge machine	142
5.3	Conventional die sinking EDM	143
5.4	(a) Wire EDM, (b) micro EDM mill	144
5.5	Tools of different cross sections and produced workpiece	144
5.6	Some applications of EDM machining	145
5.7	Surface roughness parameter (R_a)	147
5.8	Procedure for response surface methodology (RSM)	151
5.9	Residual graphs for material removal rate	154
5.10	Residual graphs for surface roughness	155
5.11	Microstructure of the as-produced composite (Kanagarajan *et al.*, 2008)	157
5.12	(a) Effect of electrode rotation and pulse on-time on MRR, (b) effect of peak current and pulse on-time on MRR	158
5.13	(a) Effect of flushing pressure and pulse on-time on MRR, (b) effect of peak current and electrode rotation on MRR	159
5.14	(a) Effect of flushing pressure and electrode rotation on MRR, (b) effect of flushing pressure and peak current on MRR	160
5.15	Micrograph of the machined workpiece with heat affected zone and recast layer	161
5.16	(a) Effect of electrode rotation and pulse on-time on surface roughness, (b) effect of peak current and pulse on-time on surface roughness	161

5.17 (a) Effect of flushing pressure and electrode rotation on surface roughness, (b) effect of flushing pressure and peak current on surface roughness 162
5.18 (a) Effect of flushing pressure and pulse on-time on surface roughness, (b) effect of peak current and electrode rotation on surface roughness 163
5.19 Surface produced on machining of WC/Co composites 164
6.1 Classification of MMCs based on matrix material 172
6.2 Classification of primary processing of MMCs 173
6.3 Classification of secondary processing of MMCs 175
6.4 Tool materials used for drilling of MMCs 177
6.5 Research studies conducted on unconventional machining of various materials 184
6.6 Research studies conducted on various unconventional hole making processes 185
6.7 Schematic diagram of ultrasonic set-up 186
6.8 Basic scheme of electrical discharge machining 189
6.9 Basic scheme of electrical discharge drilling 189
7.1 Schematic representation of (a) the anodic dissolution of metals and (b) electrochemical machining with a contoured tool electrode 196
7.2 The laboratory machine for μECM as designed and fabricated by us. (a) Photograph of the equipment and (b) schematic representation of the main components 200
7.3 Tapered contour of the micro-holes utilized in the definition of dimensional accuracy (DA) 203
7.4 Influence of the amplitude of the pulses of electric current (frequency of 100 kHz) on (a) metal removal rate (MRR), (b) dimensional accuracy (DA) and (c) surface quality of the micro-holes. The electrolyte concentration is 60 mL/L 204

Published by Woodhead Publishing Limited, 2013

7.5 Influence of the frequency of electric current pulses (amplitude of 2 A) on (a) metal removal rate (MRR), (b) dimensional accuracy (DA) and (c) surface quality of micro-holes. The electrolyte concentration is 30 mL/L 206
7.6 Acquired signals of current and voltage in the oscilloscope for pulses of electric current with an amplitude of 2 A and a frequency of (a) 1 MHz and (b) 100 kHz. The electrolyte concentration is 30 mL/L 207
7.7 Influence of electrolyte concentration on (a) material removal rate (MRR), (b) dimensional accuracy (DA) and (c) surface quality of micro-holes for pulses of electric current (amplitude 2 A, frequency 100 kHz) 208
8.1 Schematic representation of: (a) metal cutting, (b) blanking and shearing, (c) forging and (d) strain-rate vs. strain loading paths 214
8.2 Cam-driven electromagnetic mechanical testing machine. (a) Photograph of the testing machine; (b) detail of the fixed housing; (c) schematic drawing showing the translating cam, follower, compression platens, load cell and displacement sensor 216
8.3 Testing machine equipped with a logistic cam. (a) Schematic representation of the cam profile and follower; (b) photograph of the logistic cam; (c) cam profile and pressure angle: (d) velocity of the ram v_x^{avg} =10 m/s in the working region of the cam follower 218
8.4 Kinematics of the testing machine equipped with a logistic cam. (a) Velocity and acceleration of the follower; (b) jerk in the working region 220

8.5 Testing machine equipped with a root-type cam. (a) Photograph of the root-type cam; (b) cam profile and pressure angle; (c) velocity and acceleration of the follower in the working region 220

8.6 Electromagnetic actuator showing the ram, coils, half-ring inserts and ferromagnetic spacers 222

8.7 Experimental evolution of the force vs. displacement for logistic and root-type cam profiles in case of the mechanical testing of aluminium AA1050-O 224

8.8 Mechanical testing of aluminium AA1050-O using a logistic cam. (a) Strain-rate vs. strain loading paths; (b) material stress response with respect to strain and strain-rate (experimental data and fitting) 225

8.9 Mechanical testing of aluminium AA1050-O using a root-type cam. (a) Strain-rate vs. strain loading paths; (b) material stress response with respect to strain and strain-rate (experimental data and fitting) 226

8.10 Experimental stress–strain curves for aluminium AA1050-O for different values of strain-rate 226

8.11 Material stress response of technically pure lead with respect to strain and strain-rate (experimental data and fitting) using (a) a logistic cam and (b) a root type cam 227

8.12 Finite-element modelling of orthogonal metal cutting. (a) Mesh and detail of the plastically deforming region; (b) computed strain-rate vs. strain loading paths for different flow routes along the plastically deforming region 228

8.13 Finite-element modelling of double-notched cylindrical test specimens loaded in shear. (a) Mesh and detail of the plastically deforming region; (b) computed strain-rate vs. strain loading paths for two different elements located in plastically deforming region 229

Tables

1.1 Diamond-coated cutting inserts with different coating thickness used in the experiment with different machining parameters (1 mm depth of cut for all cases) 7

2.1 Chemical composition and mechanical properties of AISI 303 44

2.2 Cutting parameters used in validation tests 50

2.3 Input parameters for simulation with ThirdWave AdvantEgde 52

2.4 Cutting conditions used in the simulation tests with ThirdWave AdvantEdge 60

2.5 Cutting conditions used in cylindering tests 62

2.6 Arithmetic average surface roughness (R_a) and maximum peak to valley height (R_t) for different cutting speeds (bold: minimum values: underlined: maximum values) 66

2.7 Work hardening measures carried out in the transverse section of specimens machined with different cutting speeds 72

2.8 Chip thickness (h_c), chip compression ratio (r_c) and shearing angle (ϕ) at different cutting speeds 85

3.1	Experimental factor levels employed for the full factorial experimental design	103
5.1	Process parameters and their levels used for the investigation	148
5.2	Experimental results	148
5.3	Analysis of variance for material removal rate	152
5.4	Result of ANOVA response surface model for surface roughness	153
6.1	General comparative study of the different hole making processes for various materials	174
6.2	Classification of unconventional machining processes based on the energy scheme used for material removal	183
7.1	The plan of experiments	203
8.1	The plan of experiments for aluminium AA1050-O	222
8.2	The plan of experiments for technically pure lead (99.9%)	223

List of abbreviations

3-D	three-dimensional
AE	acoustic emission
BUE	built-up edge
BUL	built-up layer
CBN	cubic boron nitride
CVD	chemical vapor deposition
DA	dimensional accuracy
ECM	electrochemical machining
μECM	micro-electrochemical machining
EDD	electro-discharge drilling
EDM	electrical discharge machining
EDX	energy-dispersive X-ray spectroscopy
FCC	face-centered cubic
FFT	fast Fourier transform
HSM	high-speed machining
HSS	high-speed steel
ISSF	International Stainless Steels Forum
MMC	metal matrix composite

MRR	material removal rate
MTL	material transfer layer
MQL	minimum quantity lubrication
PCD	polycrystalline diamond
PTFE	polytetrafluoroethylene
PVD	physical vapor deposition
RMS	root-mean-squared
RSM	response surface methodology
SEM	scanning electron microscopy
STFT	short-time Fourier transform
USM	ultrasonic machining
WC	tungsten carbide
WRW	workpiece removal weight

Preface

Machining and machine-tools is an important subject with application in several industries. Parts manufactured by other processes often require further machining operations before the final product is made. Traditional machining is the broad term used to describe removal of material from a workpiece, and covers chip formation operations – turning, milling, drilling and grinding, for example. Recently, the industrial utilization of non-traditional machining processes such as EDM (electrical discharge machining), ECM (electrochemical machining) and USM (ultrasonic machining) has increased. Also, the performance characteristics of machine-tools and the significant development of existing and new processes and machines need to be considered. In Europe, the USA, Japan and countries of emerging economies machine-tools is a sector undergoing great technological evolution.

Machining processes can be applied to work metallic materials, polymers, wood, ceramics, composites, biomaterials, nanomaterials and other specialist materials.

This book aims to provide information on machining and machine-tools for modern industry. Chapter 1 provides an analysis of the evolution of acoustic emission signals for monitoring diamond-coated tool delamination wear in machining. Chapter 2 is dedicated to high-performance machining of austenitic stainless steels.

Chapter 2 presents an analysis of forces monitoring in shape grinding processing of complex parts. Chapter 4 covers optimization of minimum quantity lubrication in grinding with CBN (cubic boron nitride) wheels. Chapter 5 is dedicated to EDM, in particular a study on machining characteristics of WC-Co composites.

Chapter 6 provides information on conventional and unconventional hole making in metal matrix composites (MMCs). Chapter 7 describes a laboratory machine for microelectrochemical machining. Finally, Chapter 8 is dedicated to a cam-driven electromagnetic mechanical testing machine.

The book can be used as a research tool for a final undergraduate engineering course or as a topic on machining and machine-tools at the postgraduate level. It can also serve as a useful reference for academics, manufacturing researchers, mechanical, manufacturing, industrial and materials engineers, and professionals in machining and related industries.

I would like to thank Woodhead Publishing for this opportunity and for their enthusiastic and professional support. Finally, I would like to thank all the chapter authors for their input.

J. Paulo Davim
Aveiro, Portugal
November 2012

About the contributors

Editor

J. Paulo Davim received his PhD in mechanical engineering from the University of Porto in 1997 and the Aggregation from the University of Coimbra in 2005. He is at present Aggregate Professor in the Department of Mechanical Engineering of the University of Aveiro and Head of MACTRIB – Machining and Tribology Research Group. He has more than 25 years of teaching and research experience in manufacturing, materials and mechanical engineering with special emphasis in machining and tribology. He is the editor of five international journals, and guest editor, editorial board member, reviewer and scientific advisor for many international journals and conferences. He has also published more than 40 book chapters and 350 articles as author and co-author in refereed international journals (more than 170 cited ISI Web of Knowledge, h-index = 25+) and conferences.

Authors

Chapter 1

P. Lu is currently a graduate assistant, pursuing his doctoral degree, in the Mechanical Engineering Department at The University of Alabama (UA). In 1997, he was admitted to Wuhan University of Technology, China, where he received his bachelor's degree in 2001 and master's degree in 2004. To advance his research, Mr Lu entered UA in 2009, and since then, his work has been focused on the analysis of acoustic emission signals for tool condition monitoring, indentation and sliding simulations for diamond-coated carbides, and tool wear of diamond-coated cutting tools. He was awarded a graduate council fellowship from UA in 2011–2012. Mr Lu has conducted joint research at General Motors R&D Center over two summers.

Y. K. Chou is a Professor of Mechanical Engineering at The University Alabama. He received his PhD in Industrial Engineering from Purdue University. Professor Chou's research/teaching interests include design and manufacturing, CAD/CAM, solid mechanics, tribology, metrology and materials. His recent research thrust includes diamond-coated cutting tools, electron beam additive fabrications and nanotechnology-enabled prosthesis supported by the US government and industry. Professor Chou is an active member of ASME, SME and ASM-International. He currently serves as the Chair of the Manufacturing Processes Technical Committee in ASME's Manufacturing Engineering Division; he also serves on the Scientific Committee of North American Manufacturing Research Institute of SME. Professor Chou is a fellow of ASME. He is a registered Professional Engineer in the State of Alabama.

Chapter 2

Ana Isabel Fernández Abia is associate professor of manufacturing engineering at the University of León (Spain). She has been working since 1996 in stainless steel machining, conducting studies on machinability and tool wear in turning operations of these materials. Her PhD degree was reached with work on high-performance turning of austenitic stainless steels. She participates in research projects with companies involved in manufacturing engineering.

Joaquín Barreiro García is associate professor of manufacturing engineering at the University of León (Spain). Prior to joining, he enjoyed two years researching at the University of Cantabria (Spain). His area of expertise is in manufacturing systems and processes, with particular emphasis on metal machining monitoring and inspection systems. He received his PhD degree from Polytechnic University of Madrid in 2001. He is a Fellow Member of SME and was Secretary of the Spanish Society of Manufacturing Engineering from 2006 to 2012.

Luis Norberto López de Lacalle is the head of High Performance Machining research in the Department of Mechanical Engineering of the Faculty of Engineering of Bilbao. He was born in 1964, finishing his engineering studies in 1988. His PhD was awarded in 1993 following work on reliability updating using experimental measurements of natural frequencies of mechanical systems. He then began research into the milling and turning of titanium and nickel alloys, using high-pressure coolant-assisted systems in turning and drilling operations. Other important projects were the Introduction of High Speed Machining in Die and Mould Basque Companies. His current research is looking at five-axis machining and simulation of ball-end milling operations using semi-empirical models. He is the author of more than 90 journal articles on machining issues.

Chapter 3

Antoniomaria Di Ilio graduated in Mechanical Engineering in 1977 at the University of Rome and began his academic career at the University of Naples, undertaking research in the field of composite materials. He has been professor of manufacturing technology at the University of L'Aquila since 1989, where he is Director of the Degree Course in

Engineering Management. He is the author of more than 90 scientific papers concerning the manufacturing technologies of metals and advanced materials, with traditional as well as non-traditional methods. In particular, he has performed research on grinding and drilling of composite materials as well as in the field of laser technologies. In 2012 he was appointed Associate Member of the International Academy for Production Engineering (CIRP).

Chapter 4

Eduardo Carlos Bianchi is a Professor in the Department of Mechanical Engineering, School of Engineering, University Estadual Paulista (Unesp), Bauru Campus. He graduated in Mechanical Engineering at Unesp-Bauru, and obtained his Master of Science and PhD degrees in Mechanical Engineering at the University of Sao Paulo (USP), in the areas of Manufacturing Processes and Abrasive Machining. He is a founder and Senior Member of the Grinding Research Group, leading the Laboratory of Abrasive Machining (LUA). He teaches undergraduate and graduate courses, mostly in Manufacturing Technology and Abrasive Machining Processes. He is a member of the Manufacturing committee of the Brazilian Society of Mechanical Sciences and Engineering (ABCM), and member of the Brazilian Association of Technical Standards (ABNT). He has advised on seven masters dissertations and three doctoral theses, and has published more than 100 papers in national and international journals.

Anselmo Eduardo Diniz graduated in Mechanical Engineering in 1982 from the University of Campinas (UNICAMP), Brazil. He received his doctorate from UNICAMP in 1989. In 1990–1991 he undertook postdoctoral studies at the University of California at Berkeley (CA-USA) under the supervision of Prof. David Dornfeld. He has been a professor of the Mechanical Engineering course of UNICAMP since 1983, teaching courses in the area of Manufacturing, mainly in Metal Cutting. Since 1990 he has been teaching graduate courses. He has advised 34 master dissertations and 13 doctoral theses. He is the author of 36 papers published in journals cited in the ISI Web of Knowledge database and more than 100 conference papers. He also works as referee for several international journals. From 2006 to 2010 he was the director of the College of Mechanical Engineering of UNICAMP.

Paulo R. Aguiar is a Professor in the Department of Electrical Engineering, School of Engineering, Unesp, Bauru Campus. He was a postdoctoral fellow at the University of Connecticut, USA, during 1998–1999, in the area of Manufacturing Processes. After receiving a Master of Science and PhD in Electrical Engineering from Sao Paulo University, USP (Electrical Machine, Power Systems and Manufacturing Processes), Dr Aguiar took the position of Associate Professor in 2003 at Unesp-Bauru, through a public examination. He has lectured on numerous courses in the areas of Control Systems, Electrical Machines and Manufacturing Processes. He received an Electrical Engineering degree in 1988 at Unesp-Bauru. He has been supported by private industry, and state and federal research agencies since graduating. He is a founder and Senior Member of the Grinding Research Group, leading the Laboratory of Abrasive Machining (LUA) and the Data Acquisition and Signal Processing Laboratory (LADAPS) at Unesp.

Rubens Chinali Canarim is a Mechanical Engineer at the Department of Mechanical Engineering, School of Engineering, University Estadual Paulista (Unesp), Bauru Campus. He is a graduate student in Mechanical Engineering under the supervision of Prof. Dr Eduardo Carlos Bianchi, and is a member of the Grinding Research Group. He has also worked as Substitute Professor at Unesp, teaching courses in Fluid Mechanics and Heat and Mass Transfer.

Chapter 5

K. Palanikumar received his PhD in Mechanical Engineering from Anna University, Chennai, India. He is currently Professor and Principal in Sri Sai Ram Institute of Technology, Chennai, Tamilnadu, India. He has published more than 75 international journal articles and more than 100 national journal and conference articles. He is on the Review Board of many international journals in Materials and Manufacturing. He is on the Editorial Board of two international journals. His current research interest includes machining of composite materials, welding of metals and alloys, total quality management and computer-integrated manufacturing.

J. Paulo Davim, as above.

Chapter 6

Inderdeep Singh is Assistant Professor at the Department of Mechanical and Industrial Engineering, Indian Institute of Technology, Roorkee. He received his PhD degree from the Indian Institute of Technology (IIT), Delhi. His research work won an award from the Foundation for Innovation and Technology Transfer (FITT) as the 'Best Industry Relevant PhD Thesis of the Academic Year 2004–2005' at IIT Delhi. He is involved in research and development work in the field of machining of composite materials. He has guided two PhD theses and is currently supervising six PhD scholars. He has also supervised 20 postgraduate students towards successful completion of their dissertation work. He has to his credit 100 research publications in international and national journals and conference proceedings.

Sarbjit Singh is working as Assistant Professor at PEC University of Technology, formally known as Punjab Engineering College Chandigarh, India. He received his Bachelors and Masters in Mechanical Engineering and in Production Engineering, respectively, from Punjab Technical University. He has more than 10 years of teaching experience in the field of Manufacturing Sciences, Unconventional and Conventional Machining Processes, and Engineering Graphics. He is on deputation at the Indian Institute of Technology, Roorkee, to pursue his doctorate in the field of primary and secondary processing of metal matrix composites. He has published ten research papers in international journals and proceedings of international and national conferences.

Abhishek Singh is Research Scholar at the Department of Mechanical and Industrial Engineering, Indian Institute of Technology, Roorkee. He completed his M-Tech in Manufacturing Technology at the Department of Mechanical Engineering, National Institute of Technical Teacher's Training and Research Kolkata, India. He is involved in research and development in the field of metal matrix composites and hybrid machining. His fields of interest include process modelling and optimization, advanced manufacturing processes and metal matrix composites. He has published five research papers in international journals and proceedings of international and national conferences.

Chapter 7

Paulo A. F. Martins received his PhD in Mechanical Engineering from Instituto Superior Técnico (IST), Technical University of Lisbon, Portugal, in 1991 and received his Habilitation in 1999 in recognition of his work in the numerical and experimental simulation of metal forming processes. He is currently Professor of Manufacturing at IST. His research interests include metal forming and metal cutting and he has co-authored four books, several national and international patents, and 250 papers in international journals and conferences. He is associate member of CIRP (The International Academy for Production Engineering) and belongs to the editorial board of several international journals.

Pedro A. R. Rosa received his PhD in mechanical engineering from Instituto Superior Técnico (IST), Technical University of Lisbon, Portugal, in 2007. He is currently Assistant Professor of Manufacturing at IST. His research interests include metal cutting and micro manufacturing and he has co-authored one book, five book chapters, and 35 papers in international journals and conferences. He is member of the Portuguese Associated Laboratory for Energy, Transports and Aeronautics and head of the Machining and Micro Manufacturing Research Group.

Ivo M. F. Bragança received his MSc in Mechanical Engineering from Instituto Superior Técnico (IST), Technical University of Lisbon, Portugal, in 2008. He has been conducting PhD research work at IST since 2009. His research interests include machine-tools, non-traditional processes and electro discharge machining, and he is currently a Teaching Assistant of Micro Manufacturing at IST. He is the co-author of seven papers in international journals and conferences, and collaborator of the Portuguese Associated Laboratory for Energy, Transports and Aeronautics.

Gabriel R. Ribeiro received his BSc in Biotechnology Engineering in 2005 from UAlgarve and his MSc in Mechanical Engineering from Instituto Superior Técnico (IST), Technical University of Lisbon, Portugal, in 2009. He has been conducting PhD research work at IST since 2010. His research interests include machine-tools, non-traditional processes and electrochemical machining, and he is currently Teaching Assistant of Micro Manufacturing at IST. He is the co-author of five international papers in international journals and conferences, and a collaborator of the Portuguese Associated Laboratory for Energy, Transports and Aeronautics.

Chapter 8

Paulo A. F. Martins, as above.

Pedro A. R. Rosa, as above.

Carlos M. A. Silva received his MSc in Mechanical Engineering from Instituto Superior Técnico (IST), Technical University of Lisbon, Portugal, in 2007. He has been conducting PhD research work at IST since 2009. His research interests include mechanical and fracture characterization of materials, testing machines, metal cutting and metal forming and he is currently Teaching Assistant of Machine-Tools at IST. He is the co-author of seven papers in international journals and conferences, and collaborator of the Portuguese Associated Laboratory for Energy, Transports and Aeronautics.

1

Analysis of acoustic emission signal evolution for monitoring diamond-coated tool delamination wear in machining

P. Lu[1] and Y. K. Chou[1]

Abstract: Diamond-coated cutting tools have been utilized as a cost-effective alternative to brazed polycrystalline diamond tools in applications such as machining lightweight high-strength materials (e.g. metal matrix composites). However, coating delamination is a major failure mode of diamond-coated tools, terminating tool life prematurely. Once delamination failure occurs, the tool substrate often suffers severe abrasive wear, leading to catastrophic tool failures thus interrupting machining operations. Accurate detection and forecasting of coating delamination events may thus prevent production losses and assist process planning. In this study, the characteristics of acoustic emission (AE) signals acquired during machining of an aluminum matrix composite using diamond-coated cutting tools were analyzed in various ways. The AE signals were analyzed in both the time and the frequency domains under various machining conditions and at different cutting times. The results from machining experiments and analysis indicate that it may be feasible to use AE signals to monitor the condition

[1] Mechanical Engineering Department, The University of Alabama, USA.

Published by Woodhead Publishing Limited, 2013

of diamond-coated tools in machining. AE root-mean-squared values decrease considerably once coating delamination occurs. The results also indicate a correlation between tool condition and the fast Fourier transformation (FFT) spectra of AE raw data. The AE-FFT spectra with cutting time generally show a decreased intensity for the low-frequency peaks, but increased intensity for the high-frequency peaks. In addition, AE-FFT analysis of data from various time periods during one cutting pass clearly indicate coating failure transition. Further research using the short-time Fourier transform (STFT) method shows that during the coating failure pass, there is a clear increase in the amplitude ratio (1/value change) of the high- vs. low-frequency component with cutting time, which captures the coating failure transition. Repeatable results indicate that the applied STFT method has the potential for monitoring of diamond-coated tool failure during machining. However, for coating failures associated with less tool wear (flank wear-land width <0.8 mm), the amplitude ratio plot from the STFT analysis may not clearly identify the failure transition.

Key words: Acoustic emission, delamination, diamond-coated tool, machining.

1.1 Introduction

Synthetic polycrystalline diamond (PCD) is commonly used in industry to machine non-ferrous materials because of its exceptional tribological properties. However, the processing and fabrication of PCD tools are costly. On the other hand, diamond-coated tools created by chemical vapor deposition (CVD) on carbide substrates have been developed and evaluated in various machining applications (Kustas *et al.*, 1997; Grzesik *et al.*, 2002), for example for machining high-strength Al–Si alloys and aluminum matrix composites. Several previous experimental investigations have shown that coating delamination is the main mechanism that dictates the life of diamond-coated cutting tools (Hu *et al.*, 2008a). Once delamination occurs, tool wear is rapid and can be catastrophic, causing parts to be rejected and possible damage to the machine tool. Moreover, delamination events are typically difficult to predict because of the complexity of the tool wear process. Thus, it is of interest for tool users to be able to detect coating delamination as a means of process monitoring. Among common sensors deployed in machining operations, acoustic emission (AE) sensors have been evaluated to monitor tool wear.

AE refers to the transient elastic waves generated during the rapid release of energy from a localised source within a solid. For metallic materials, acoustic emissions may be associated with plastic deformation, initiation and propagation of cracks, and frictional contacts between the tool, chip, workpiece, etc. Since the late 1960s, AE has been widely used to identify cracks in pressure vessels, bridges, hydroelectric dams, composite laminates, etc. (Beggan *et al.*, 1999). Among the wide range of non-destructive evaluation techniques for detecting cutting tool conditions, AE has been recognized as a feasible method for monitoring of in-process tool wear due to its sensitivity to tool wear (Iwata and Moriwaki, 1977). In machining, AE signals can be easily distinguished from signals associated with machine vibrations and ambient noise because of its high-frequency nature, for example from 10 kHz to 1 MHz (Dornfeld and Lan, 1983).

Dornfeld's group at the University of California at Berkeley have perhaps been the pioneers in the study of AE signals in cutting and exploring AE applications for monitoring of machining processes. Dornfeld and Kannatey (1980) performed orthogonal cutting tests, varied the process parameters and recorded the AE signals generated. They noted a strong dependence of the AE root-mean-squared (RMS) intensity on both the strain rate and the cutting speed. Moreover, Lan and Dornfeld (1982) reported that the AE power spectrum is of high amplitude at a specific frequency range during tool fracture. In a separate study, the same authors reported that tool fractures and catastrophic failures generate burst AE signals (Dornfeld and Lan, 1983).

It has also been observed that an abrupt transition of AE magnitudes may occur with progression of tool wear (Mukhopadhyay *et al.*, 2006). By contrast, more recently, Feng *et al.* (2009) analyzed the influence of tool wear on a microgrinding process by analyzing AE signals, and the results showed that AE-RMS signals are not monotonic with tool wear magnitude, and thus may not be suitable for monitoring of tool wear. Haber *et al.* (2004) claimed that AE-RMS signals are robust and can provide a versatile means of detecting contact between the tool and the workpiece. In addition, it was pointed out that, from spectral analysis of AE signals, AE signals are very sensitive to the tool condition changes, with increasing amplitudes at a high frequency range, up to 160 kHz. More recently, Jemielniak and Arrazola (2008) used AE sensors to monitor tool condition in micromilling and reported that AE signals still show a dependence on tool wear in microscale cutting. Kanga *et al.* (2008) also applied AE sensors for tool condition monitoring in machining of small-scale parts. The authors argued that AE-RMS values can be used for tool condition monitoring.

Investigations of AE signals for tool wear monitoring of coated tools are relatively scarce, and even fewer literature sources indicate an understanding of AE characteristics and their relationship with the wear and failure mechanisms of coated cutting tools. Somasundaram and Raman (1993) conducted machining tests with different types of coated tools and noted that AE-RMS values vary with the coating type and do not necessarily increase with tool flank wear. Moriwaki and Tobito (1990) reported that there was a transition from the burst-type to the continuous AE-RMS amplitude with the progression of tool wear. In addition, the RMS response increased in both amplitude and fluctuation as the tool was about to reach the end of its life. Cho and Komvopoulos (1997) performed fast Fourier transform (FFT) analysis of the AE signals collected when machining by coated tools. Their results indicated that coating delamination may produce high-frequency burst-type signals. Preliminary investigations by our lab (Hu *et al.*, 2008b; Lu *et al.*, 2009, 2010) applied an AE sensor to monitor coating failures of diamond-coated tools in machining. It was found that AE-RMS changes might be noticeably reduced before and after coating failure. Moreover, the frequency response of AE signals during machining may alter significantly before and after coating failure. In addition, burst signals of AE-RAW (raw) data were also noted, but not consistently. However, the results were not able to specify the failure transition period. It is possible that the diamond coating is brittle and failures may occur during entry or exit of the cutting tool, which may become difficult to identify by simply analyzing AE signals at different cutting passes. Thus, monitoring coating failures, especially the failure transition, via AE signals remains a serious challenge with regard to coated tools during machining operations. The primary objective of this study was to examine the AE signals produced when using diamond-coated tools in machining aluminum matrix composites, and to correlate AE characteristics with tool conditions, for example coating failures. This will potentially lead to more accurate detection and possibly forecasts of coating delamination events, thereby preventing production losses and assisting in process planning. The AE signals from different cutting passes in machining by diamond-coated tools are analyzed in detail, for example AE-FFT spectra in subdivided cutting zones and the amplitude ratio of low vs. high frequency by using short-time Fourier transform (STFT); the results are used to correlate the characteristics of AE signals with the tool conditions, especially coating failures.

1.2 Experimental set-up and methods

The tool substrates used to deposit diamond coatings were fine-grained tungsten carbides (WC) with 6 wt% cobalt, in the shape of square inserts (SPG422). Chemical etching pretreatment was applied to the samples with the aim to modify the as-ground surface before diamond deposition, improving adhesion of the coating by halting the effect of the cobalt binder in the cemented carbide substrate. The pretreatment conditions were selected from previous analyses. For the coating process, diamond films were deposited using a high-power microwave plasma-assisted CVD process. A mixture of methane in hydrogen was used as the feedstock gas. Nitrogen, maintained at a certain ratio to methane, was inserted to the gas mixture to obtain nanostructures by preventing cellular growth. The pressure was about 30–55 Torr and the substrate temperature was about 685–830 °C. The coating thickness varied from about 5 μm to about 25 μm, as estimated from the tool cutting edge radius measured via an interferometer.

A computer numerical control lathe, Hardinge Cobra 42, was used to perform machining experiments, outer diameter turning, to evaluate the tool wear of diamond-coated tools. The set-up is shown in Figure 1.1. With the tool holder used, the diamond-coated cutting inserts formed a 0° rake angle, 11° relief angle and 75° lead angle. The workpieces were

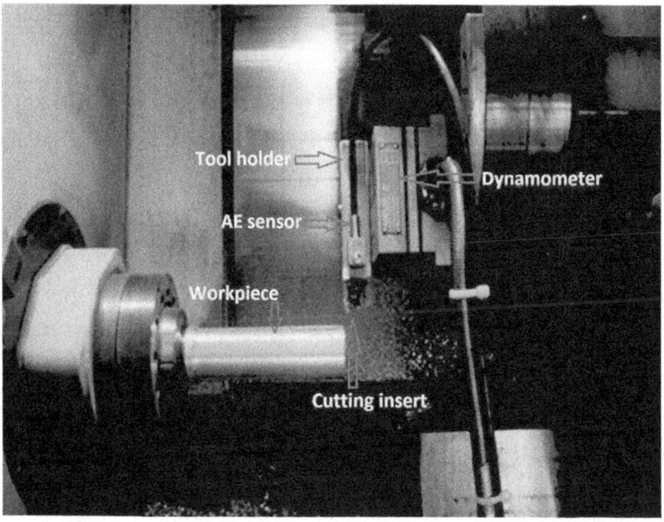

Figure 1.1 Experimental set-up in a CNC lathe showing an AE sensor mounted on the tool holder (taken from Hu *et al.*, 2008)

round bars made of A359/SiC-20p composite. The machining conditions used (e.g. cutting speed and feed) were varied for different tools used, except a fixed 1-mm depth of cut. Machining was conducted without coolant. Cutting inserts with different coating thicknesses (Table 1.1) were used in several tests to analyze AE signals collected during machining

Table 1.1 Diamond-coated cutting inserts with different coating thickness used in the experiment with different machining parameters (1 mm depth of cut for all cases)

Insert	Coating thickness (μm)	Cutting speed (m/s)	Feed (mm/rev)
A, B, C	15	4	0.3
D	5–10	4	0.15
E	25	4	0.15
F	25	8	0.3

by diamond-coated tools. During machining testing, the diamond-coated cutting inserts were inspected, after each cutting pass, by optical microscopy to examine tool wear and diamond coating conditions, failure or not. Worn tools after testing, reaching above around 0.8 mm flank wear-land width, were also observed by scanning electron microscopy (SEM) to examine tool wear patterns and sizes and to confirm coating failures. In addition, cutting forces were monitored throughout testing using a Kistler dynamometer, Model 9257B.

An AE sensor was employed to acquire data, both AE-RMS and AE-RAW, during the entire operation of machining. The sensor used was an 8152B Piezotron AE sensor from Kistler, and the signal was first fed into a coupler, Kistler 5125B, for amplification and post-processing. The time constant for the RMS values was 1.2 ms, and the calculation was conducted by the RMS converter inside the AE coupler according to Eq. 1.1. The resulting AE-RAW and AE-RMS data were digitized using an IOtech high-speed DAQ3000 PCI board at 500 kHz sampling rate per channel. In addition, MATLAB software was used to evaluate the frequency response, using FFT analysis (Eq. 1.2), with the AE raw data at different cutting time.

$$U_{\mathrm{RMS}} = \sqrt{\frac{1}{\tau}\int_0^{\tau} U_{\mathrm{AE}}{}^2 \mathrm{d}t}, \qquad [1]$$

where U_{RMS} is the voltage after the RMS converter, U_{AE} is the raw data and τ is the time constant.

$$U_{\mathrm{FFT}(k)} = abs\left(\sum_{j=1}^{N} U_{\mathrm{AE}}(j)\omega_N^{(j-1)(k-1)}\right), \qquad [2]$$

where $\omega_N = e^{\frac{-2\pi i}{N}}$, and is an Nth root of unity, $U_{\mathrm{AE}(j)}$ is the jth member of the raw data with total length N, $U_{\mathrm{FFT}(k)}$ is the voltage corresponding to the *k*th member after FFT, and *abs* returns the value of the complex modulus or magnitude.

1.2.1 AE-FFT at different cutting passes

To examine if quantitative information of AE-FFT evolution can be utilized to identify coating failures, the amplitudes associated with the low-frequency peak (about 25 kHz) and high-frequency peak (around 125–160 kHz) were recorded and analyzed for each cutting pass. Then, the amplitude of the high-frequency component and the low-frequency component at different cutting passes can be plotted and compared. Figure 1.2 shows an example of the AE-FFT intensity (low and high frequencies) from the initial cutting to the tool failure cutting pass of diamond-coated insert F. During machining using insert F, the cutting tool failed around 25 s after three cutting passes.

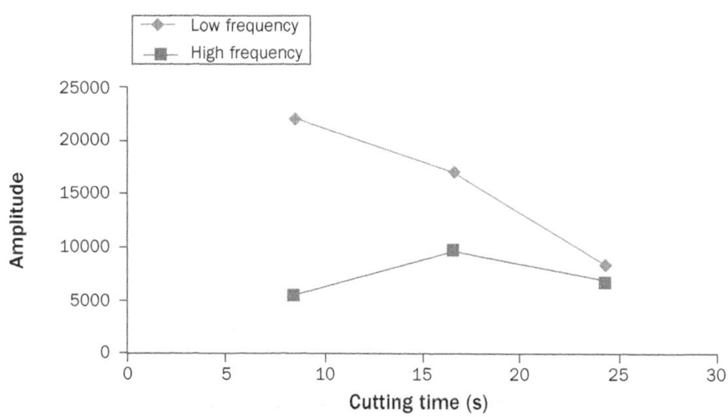

Figure 1.2 AE-FFT intensity (high and low frequencies) with cutting time for insert F

1.2.2 AE-FFT of a cutting pass with four divided subsets

The AE signals collected during the coating failure pass (confirmed by optical microscopy after machining) were further analyzed in detail; specifically, the AE-RAW data from the cutting pass were divided into four equal cutting periods and FFT analysis was further performed for each AE data subset. Figure 1.3 shows the AE-FFT spectra of insert F at each subset during the coating failure pass.

Figure 1.4 compares the AE-FFT intensity changes along the different cutting time subsets, both low and high frequencies, for insert F during the coating failure pass.

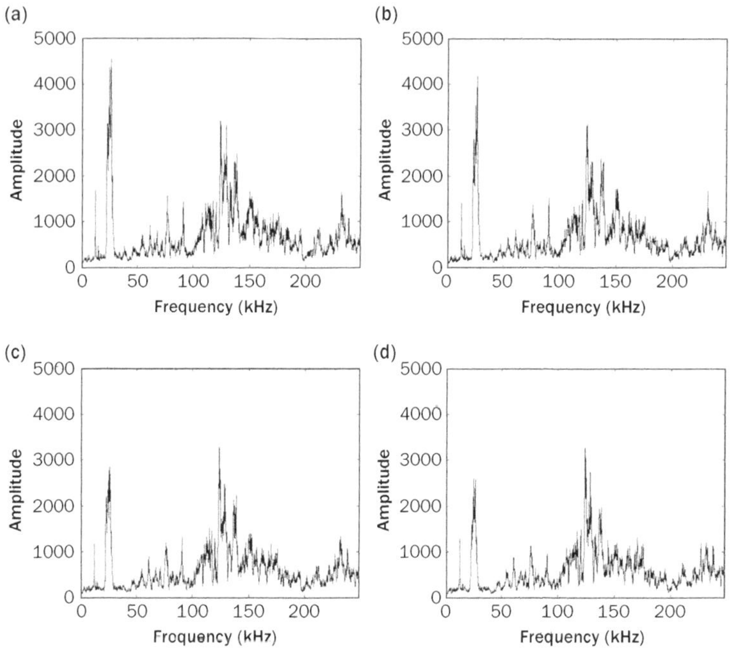

Figure 1.3 AE-FFT at different time periods for insert F during the failure pass: (a) first, (b) second, (c) third and (d) last 25 per cent cutting time of the entire machining pass

Published by Woodhead Publishing Limited, 2013

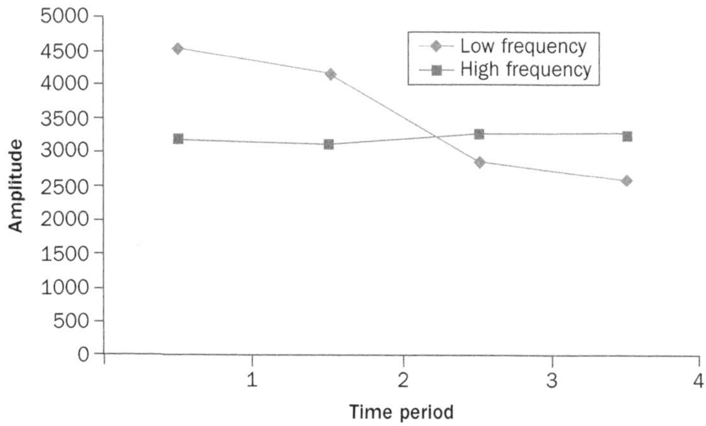

Figure 1.4 AE-FFT amplitude comparisons in the four sub-periods during the coating failure pass of insert F

1.2.3 STFT method

It was further speculated that finer cutting time subsets from a cutting pass (especially that when a significant event occurs) may be required to capture and show the coating failure transition, and thus the AE signals from different cutting passes were further analyzed based on the STFT approach. Similar methodologies have been applied in event-related desynchronization by Guilleminault (1997) and Sutoh (1997). Figure 1.5 shows a schematic of the STFT approach for one cutting pass. The procedure of this method is as follows. First, the AE-RAW data from a cutting pass are divided into many continuous subsets (e.g. *n*). Each subset has the same cutting time interval (e.g. 2 s), but overlaps with the contiguous subset, i.e. the previous subset data (e.g. subset 2), and can be 0.1 s earlier than the next subset (i.e. subset 1), where 0.1 s is the time increment. Therefore, a total of *n* data subsets can be extracted from one cutting pass. Next, each subset will be processed by the

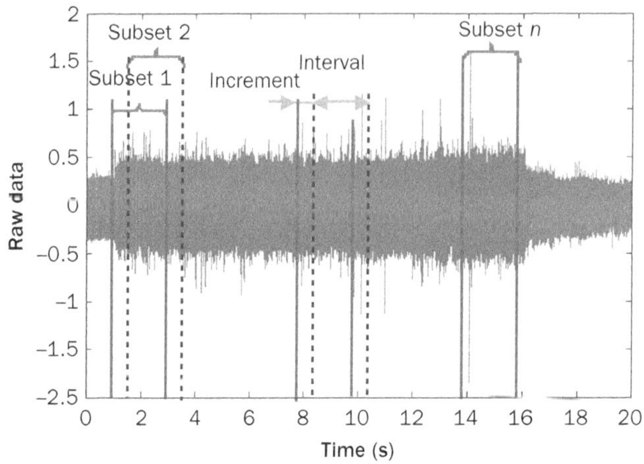

Figure 1.5 **Illustration of the short-time Fourier transform (STFT) method (from Lu *et al.*, 2010)**

FFT method, and the amplitude associated with the low-frequency peak (about 25 kHz) and high-frequency peak (around 125–160 kHz) was analyzed and recorded for each subset to compute the amplitude ratio of the high vs. low frequency. The amplitude ratio is then plotted along the cutting time of a cutting pass, to hope to capture the magnitude transition for the high-frequency component. The advantage of this method is that the transition of the coating failure can be tracked continuously to detect the event, which can be found by the change of the amplitude ratio of the high versus low frequency.

1.3 Results and discussion

In this study, machining tests were conducted first by insert F on A359 composite bars with a cutting speed of 8 m/s, a feed of 0.3 mm/rev and 1-mm depth of cut. After failure of insert F, AE signals, of different cutting passes, relating to the time domain and frequency response, including RAW, RMS and FFT, were analyzed. Machining tests were further conducted by another two diamond-coated inserts (A and B) of a different coating thickness and machining conditions (cutting speed 4 m/s, feed 0.3 mm/rev). Inserts A and B failed at around 6 and 10 min, respectively. The AE-FFT from different cutting passes was analyzed, as well as that of four subsets from different cutting passes. Moreover, the STFT method was applied to the AE signals from different cutting passes. Insert C, which had the same coating thickness as inserts A and B, was used to test the repeatability of the STFT methodology, and inserts D, E and F were used to test the STFT method under different machining conditions.

1.3.1 AE-RAW, RMS and FFT from a cutting pass

Figure 1.6 shows an example of AE-RAW signals from the initial and tool failure cutting passes for insert A. Note that the cutting tool failed around just 352 s with a large patch of coating delaminated at the tool flank surface, as observed under optical microscopy, and that AE amplitudes reduced

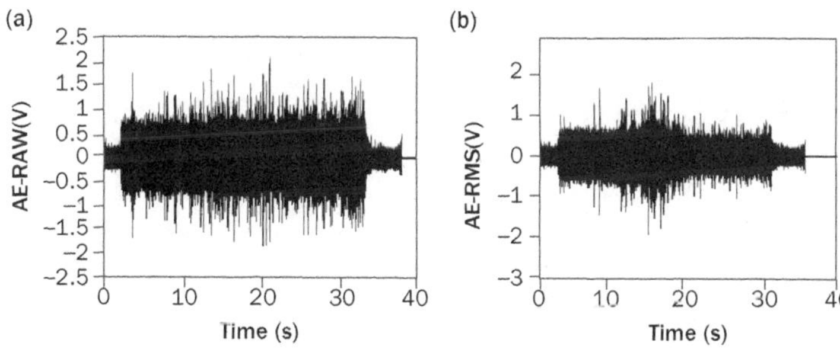

Figure 1.6 **Raw AE signal for insert A at: (a) initial cutting pass and (b) tool failure pass**

Published by Woodhead Publishing Limited, 2013

noticeably during the cut, indicating significant events had taken place in the tool and the tool–workpiece as well as tool–chip contacts.

Figure 1.7 shows overlapping AE-RMS amplitudes between the initial cutting (first cutting pass of 30 s) and the tool failure pass (failure occurred after 352 s of total cutting time), clearly showing the coating failure effects on AE-RMS amplitudes.

To investigate whether the same phenomenon could be observed for other inserts, machining tests were conducted on A359 workpieces by another diamond-coated insert (B). Figure 1.8 shows AE-RMS evolution during the coating failure pass from two identical cutting tests (same working conditions and same type of cutting inserts). One shows a clear failure transition (insert A, Figure 1.8a) while the other does not (insert B, Figure 1.8b). Therefore, AE-RMS alone from machining may not be sufficient to identify coating failure.

In addition, FFT analysis of AE-RAW signals was conducted. Figure 1.9 compares the FFT spectra from the

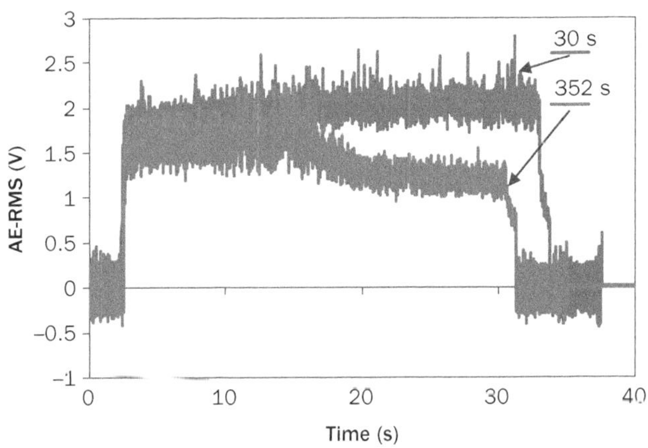

Figure 1.7 AE-RMS amplitude during initial cutting and tool failure pass for insert A (from Lu *et al.*, 2009)

Published by Woodhead Publishing Limited, 2013

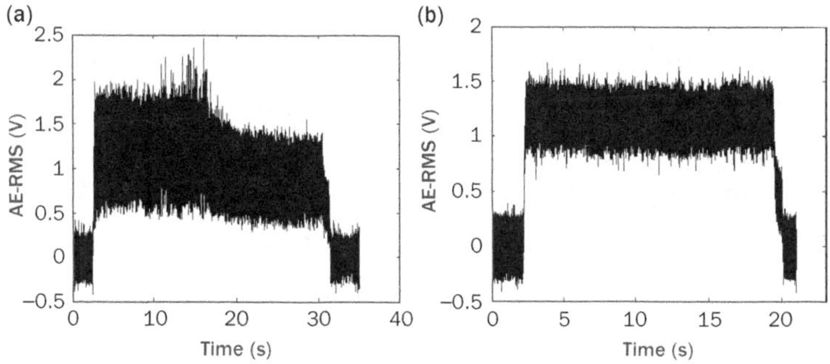

Figure 1.8 AE-RMS during tool failure pass: (a) with clear failure transition period (insert A) and (b) without clear failure period (insert B) (from Lu *et al.*, 2009)

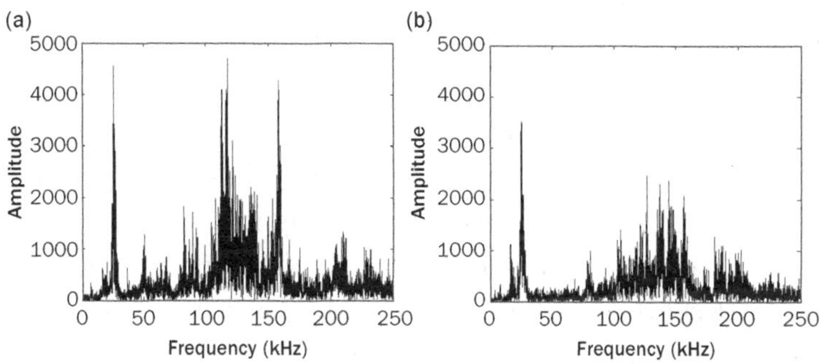

Figure 1.9 FFT spectra of raw AE signals from two cutting passes of insert A: (a) initial cutting and (b) final cutting (from Lu *et al.*, 2009)

initial cutting pass and the tool failure pass for insert A. As noted in Figure 1.9(a), i.e. initial cutting, the dominant peak appears around 25 kHz, which presumably is associated with chip formation. The remaining low-amplitude but high-frequency components were due to tool-chip/workpiece friction, and possible fractures of the reinforcement phase, SiC particles, in the work material. For the tool failure pass (Figure 1.9b), the amplitude of AE signals decreased significantly. If the diamond coating is removed from the tool substrate, chip formation and fragmentation were dramatically reduced and more energy was consumed to overcome rubbing of the substrate material against the machined surface. However, the relative intensity changed noticeably. The amplitude of low-frequency peaks (about 25 kHz) reduced with cutting time, but that of high-frequency peaks (around 125–160 kHz) increased with cutting time.

1.3.2 AE-FFT from different cutting passes

Two tests (using two inserts A and B) performed under identical cutting conditions were conducted to machine the A359 matrix composite bars. Figure 1.10(a) plots AE-FFT magnitude changes, for insert A, with cutting time for low and high frequencies, revealing a clear failure transition. By contrast, Figure 1.10(b) plots AE-FFT magnitude changes with cutting time for insert B (unclear failure transition), also for low and high frequencies. Figures 1.11 and 1.12 reveal a general trend of intensity reductions with cutting time. However, the fluctuation of the peak intensity makes specifying certain threshold values for identification of coating failure quite difficult.

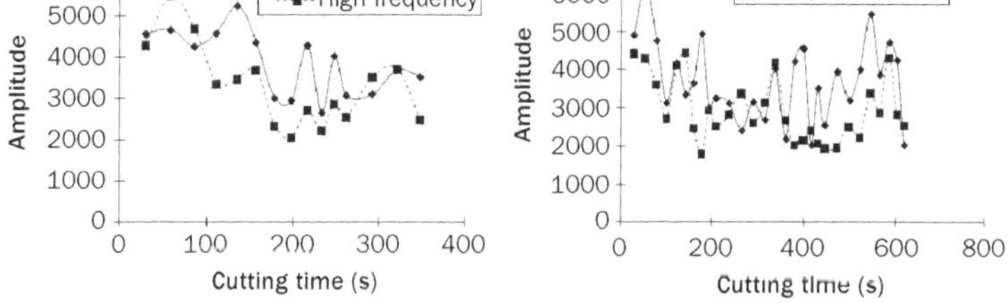

Figure 1.10 AE-FFT intensity of high- and low-frequency components with cutting time for (a) insert A and (b) insert B (from Lu *et al.*, 2009)

Published by Woodhead Publishing Limited, 2013

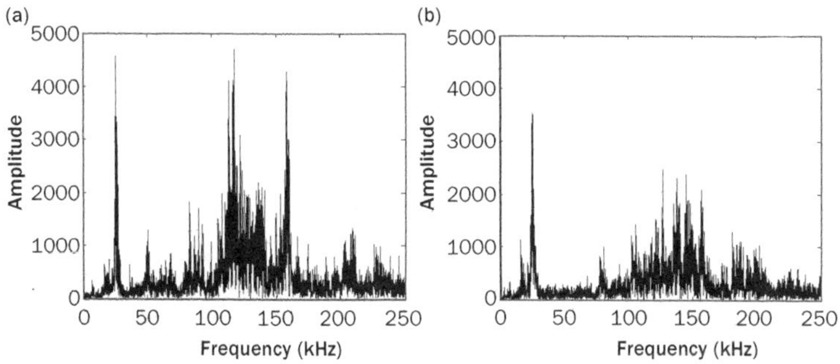

Figure 1.11 AE-FFT of insert A: (a) initial cutting pass and (b) coating failure pass

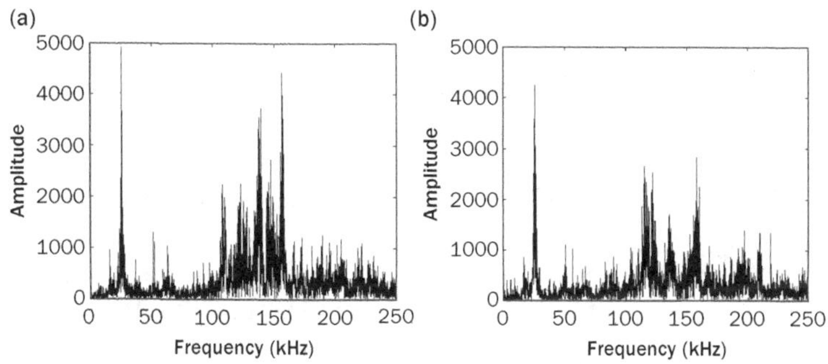

Figure 1.12 AE-FFT of insert B: (a) initial cutting pass and (b) coating failure pass (from Lu *et al.*, 2009)

1.3.3 AE-FFT of subsets of a cutting pass

The AE signals during the coating failure pass were further analyzed in detail via AE-FFT with the cutting pass divided into four subsets. Figure 1.13 compares AE-FFT spectra of insert A at different cutting periods during the coating failure pass. A clear reduction in intensity is apparent, from period 2 to periods 3 and 4.

Figure 1.14 plots AE-FFT intensity changes with cutting time, for both low and high frequencies, for two inserts

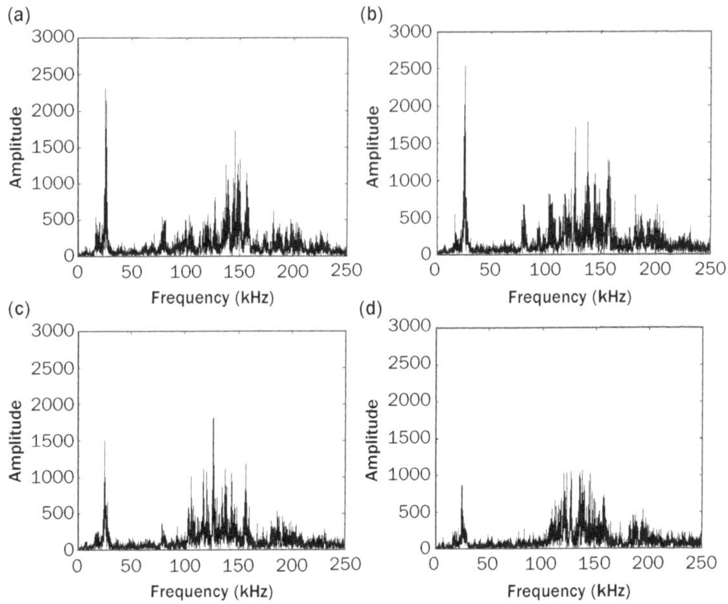

Figure 1.13 AE-FFT at different time periods for insert A during the coating failure pass: (a) first, (b) second, (c) third and (d) last 25 per cent cutting of the entire pass (from Lu *et al.*, 2009)

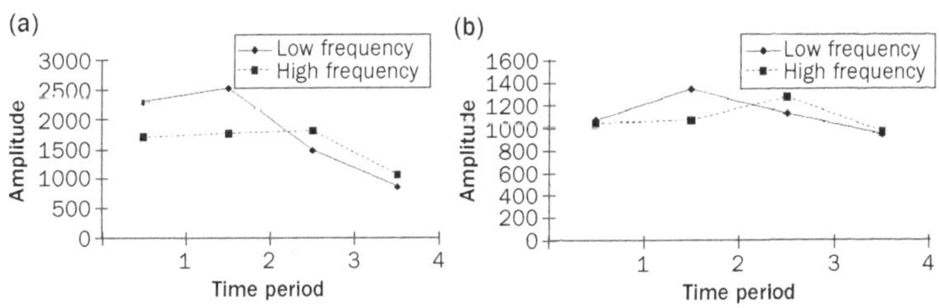

Figure 1.14 AE-FFT amplitude comparisons, between the low- and high-frequency components, in the four sub-periods during the coating failure pass: (a) insert A and (b) insert B (from Lu *et al.*, 2009)

during the coating failure pass. For insert A, between periods 2 and 3 the low-frequency peak decreases while the high-frequency peak increases. Similar results can be observed from another insert (B) between periods 2 and 3. However, the results are fairly qualitative with regard to monitoring coating failure and sometimes the difference may not be sufficiently significant to discriminate the transition. It may be difficult to identify coating failures simply by analyzing AE signals of the cutting time subsets. A more definite method based on time increments may be needed to capture the transition during the coating failure pass.

1.3.4 AE-STFT method

From the previous results, during or after the coating failure pass, the high-frequency peak will increase while the low-frequency peak will decrease. Thus, if the amplitude ratio of the high/low-frequency peak dramatically increases during a cutting pass, the coating failure pass will be identified and the transition can be captured accordingly by the STFT method. The AE-STFT method was used to analyze the AE signals from identical cutting tests for both insert A and B. Figure 1.15 first shows the amplitude ratio of the high/low-frequency peak with the STFT method of insert A at (a) initial cutting, (b) prior to failure and (c) failure. Note that the change of amplitude ratios during the coating failure pass is different from those from the other two cutting passes; a clear increase (ratio change above 1) was noted during the coating failure pass, while only a few fluctuations were observed from the other two cutting passes (much less than 1). The result from the coating failure pass of insert B (Figure 1.16), however, is different from that for insert A, without any noticeable continued increase.

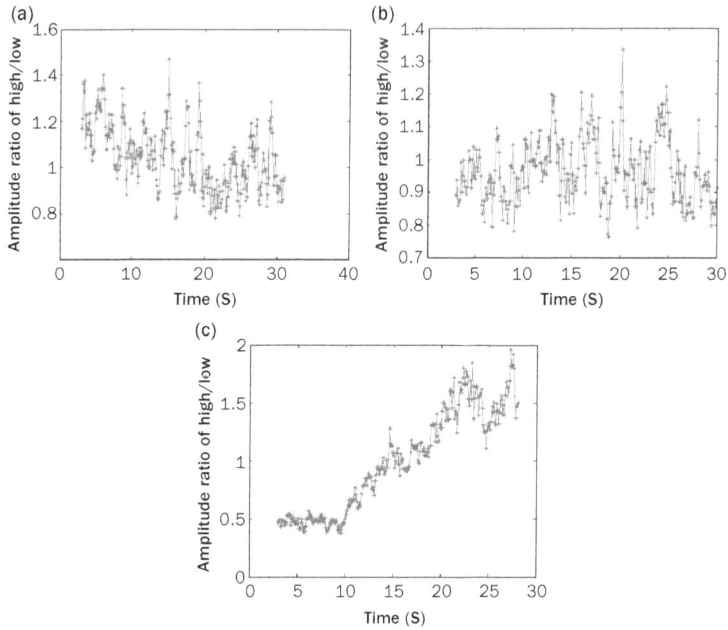

Figure 1.15 Amplitude ratio of the high/low-frequency components by the STFT method during different cutting passes of insert A: (a) initial cutting passes, (b) prior to failure pass and (c) failure pass (from Lu *et al.*, 2010)

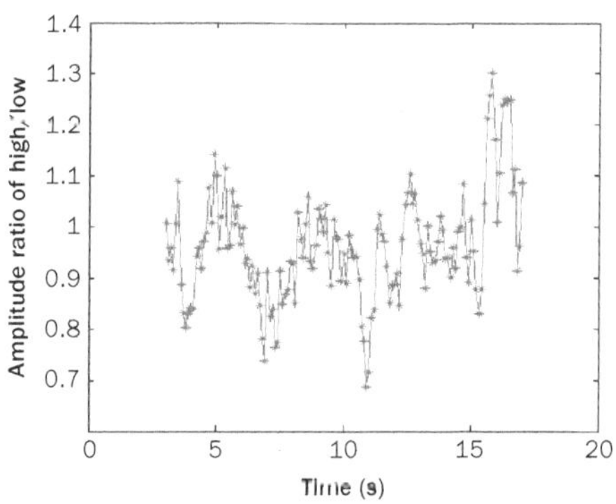

Figure 1.16 Amplitude ratio of high/low-frequency components by the STFT method during coating failure passes of insert B (from Lu *et al.*, 2010)

By further examining the value of tool wear (Figure 1.17), it was found the flank wear-land width (VB) of insert B was much smaller than that of insert A, about 0.6 mm vs. 1.7 mm. Thus, it is assumed that a certain tool wear size may exist below which actual coating delamination or failure cannot be identified via the STFT method. If further machining was conducted using this insert, the amplitude ratio of high/low-frequency peaks would reproduce the clear increase seen during the next cutting pass.

1.3.4.1 Repeatability test (insert C)

To test the assumption above with the AE-STFT method, a cutting test under identical conditions was conducted on A359 matrix composite bars using insert C (same coating thickness as inserts A and B, Table 1.1). Insert C failed after about 14 min of machining time. Figure 1.17 shows tool wear, flank wear-land width (VB), with cutting time of three inserts (A, B and C). Noticeable variation was observed. In general, the tools showed a gradual increase in wear followed by an abrupt increase in wear-land size after one or two passes. We conclude that, during these passes, coating delamination occurred and resulted in rapid wear of the

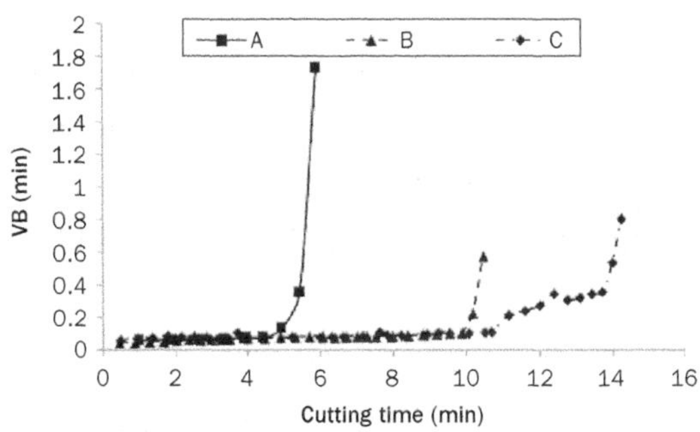

Figure 1.17 Tool wear development of cutting inserts (A, B and C) with cutting time

Published by Woodhead Publishing Limited, 2013

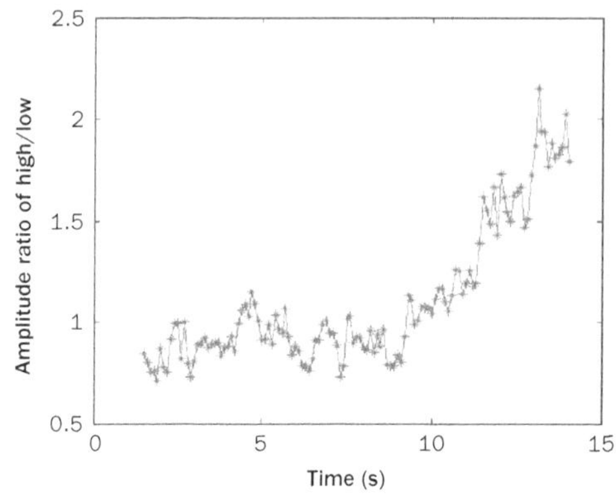

Figure 1.18 **Amplitude ratios of high/low-frequency components by the STFT method during coating failure pass for insert C (from Lu *et al.*, 2010)**

exposed substrate. Such tool wear behavior of diamond-coated tools can be consistently observed, as seen from Figure 1.17, where the tool wear of inserts A, B and C showed a gradual increase followed by an abrupt increase.

Figure 1.18 plots the amplitude ratio of high/low-frequency peaks vs. cutting time for insert C during the coating failure pass. It is clear that the change of the high/low-frequency amplitude ratio during the coating failure pass is variable, with a sharp increase (value changes over 1.5). The result is similar to that seen for insert A during the coating failure pass.

We conclude from the results above that if the flank wear of a diamond-coated insert exceeds a certain limit (e.g. 0.8 mm VB here), there will be a sharp increase in the high/low-frequency amplitude ratio during the coating failure pass.

1.3.4.2 Additional testing (inserts D, E and F)

To further examine the hypothesis that a sharp increase in the high/low-frequency amplitude ratio, from the STFT method, will occur during the coating failure pass if the flank wear of an insert exceeds a certain limit, other diamond-coated inserts (D, E and F) were used to machine the A359 matrix composites under different conditions (coating thickness and machining conditions). Figure 1.19 shows tool wear (VB) versus cutting time of inserts D, E and F, which had a coating thickness different from those of inserts A, B and C. In these tests, inserts D and E had the same machining parameters as inserts A, B and C for machining the A359 matrix composites. The VB values of the inserts all exceeded 0.8 mm after the final cutting pass.

Figure 1.20 shows the high/low-frequency amplitude ratio obtained by the STFT method during the coating failure pass for inserts D, E and F. The same sharp ratio increase during the coating failure pass was noted for all three inserts, except that insert E exhibited a decrease followed by a notable increase.

Based upon the results obtained, it can be confirmed that the amplitude ratio of the high-frequency component (about 125–160 kHz) to the low-frequency component (about 25 kHz) may be used to monitor and capture the coating failure event using the STFT of the AE raw signals.

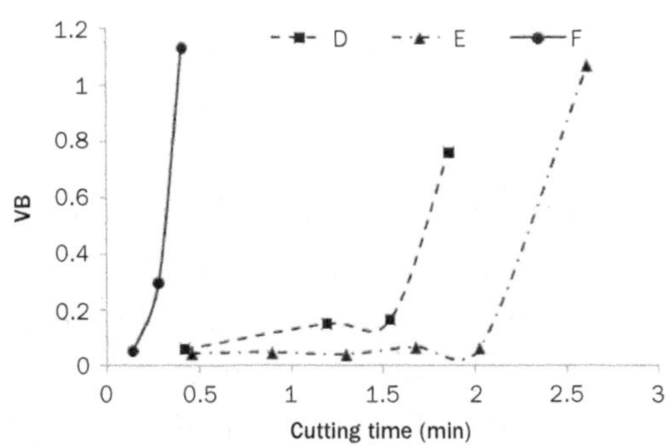

Figure 1.19 **Tool wear development of inserts D, E and F with cutting time (from Lu *et al.*, 2010)**

Published by Woodhead Publishing Limited, 2013

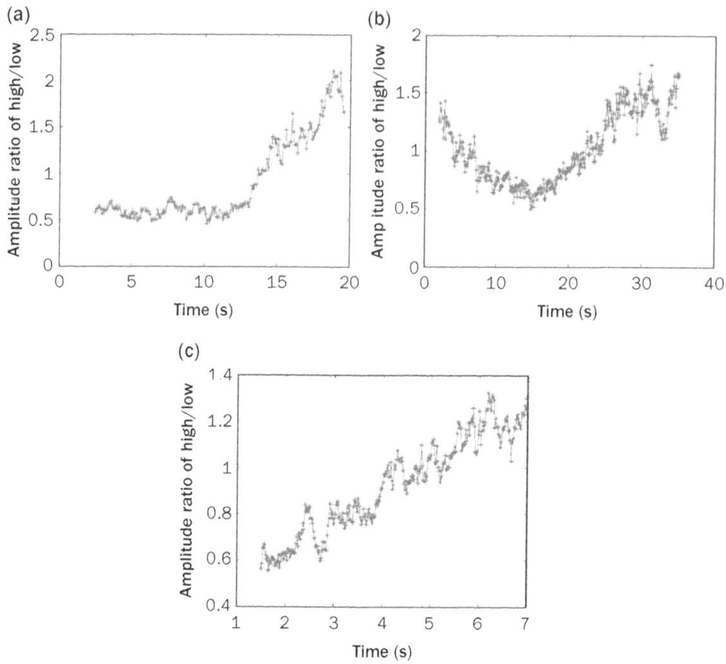

Figure 1.20 **Amplitude ratio of high/low-frequency components during the coating failure pass: (a) insert D, (b) insert E and (c) insert F (from Lu *et al.*, 2010).**

1.4 Conclusions

Coating delamination is the primary failure mechanism of diamond-coated cutting tools in machining, and from a process monitoring viewpoint, it is necessary to accurately capture the coating delamination of diamond-coated cutting tools during machining operations. In this study, cutting experiments were designed and conducted to investigate AE signals generated during machining of A359/SiC/20p composite bars using diamond-coated cutting tools. AE signals were collected and analyzed thoroughly, in particular the frequency response during a cutting pass as well as along subdivisions of the cutting pass. The intent was to correlate the characteristics of the AE spectral components with coating delamination. The main results can be summarized as follows.

1. AE-RMS values decrease significantly after coating delamination in a diamond-coated cutting tool during machining of A359/SiC-20p composites. However, in some cases, these may not show a clear transition linked to coating delamination.
2. Frequency analysis of AE signals indicates the transitional nature of the spectrum in terms of both amplitude and peak distributions. This implies that the frequency response of the AE signal may be a suitable technique to monitor the condition of diamond-coated cutting tools in machining.
3. The FFT spectra of AE data with cutting time generally show a decreased intensity for the low-frequency peaks, but increased intensity for the high-frequency peaks. In addition, the AE FFT spectra of subdivided cutting time zones during one cutting pass may clearly indicate the coating failure transition. However, this is rather qualitative for monitoring of coating failures and sometimes the difference may not be significant enough to discriminate the transition.
4. During a coating failure pass, there is a clear increase of the amplitude ratio (value change over 1) of the high- vs. low-frequency component with cutting time. The STFT method has potential for monitoring of diamond coating failure in machining. However, for coating failures associated with smaller tool wear (<0.8 mm VB), the amplitude ratio plot from the STFT analysis may not clearly distinguish the failure transition.

Acknowledgments

This material is based upon work supported by the National Science Foundation (NSF) under Grant No. CMMI 0728228.

References

Beggan C, Woulfe M, Young P, Byrne, G (1999), 'Using acoustic emission to predict surface quality', *Int J Adv Manufact Technol*, **15**: 737–42.

Cho S S, Komvopoulos K (1997), 'Correlation between acoustic emission and wear of multilayer ceramic coated carbide tools', *ASME J Manufact Sci Engng*, **119**: 238–46.

Dornfeld D A, Kannatey, A E (1980), 'Acoustic emission during orthogonal metal cutting', *Int J Mech Sci*, **22**: 285–96.

Dornfeld D A, Lan M S (1983), 'Chip form detection using acoustic emission', *Proceedings of 13th North American Manufacturing Research Conference*, pp. 386–9.

Feng J, Kim, B S, Shih, A, Ni, J (2009), 'Tool wear monitoring for micro-end grinding of ceramic materials', *Journal of Materials Processing Technology*, **209(11)**: 5110–16.

Grzesik W, Zalisz Z, Nieslony, P (2002), 'Friction and wear testing of multilayer coatings on carbide substrates for dry machining applications', *Surf Coatings Technol*, **155**: 37–45.

Guilleminault C, Black J, Carrillo O (1997), 'EEG arousal and upper airway resistance syndrome', *Electroencephalogr Clin Neurophysiol*, **103**: 11.

Haber R E, Jiménez J E, Peres C R, Alique J R (2004), 'An investigation of tool wear monitoring in a high-speed machining process', *Sensors and Actuators A: Physical* **116**: 539–45.

Hu J, Chou Y K, Thompson R G, Burgess J, Street, S (2008a), 'Nanocrystalline diamond coating tools for machining high-strength Al alloys', *Int J Refractory Metals Hard Materials*, **26**: 135–44.

Hu J, Qin F, Chou Y K, Thompson, R G (2008b), 'Characteristics of acoustic emission signals in machining using diamond coated cutting tools', *Proceedings of the 2008 International Manufacturing Science and Engineering Conference*, American Society of Mechanical Engineers, New York, MSEC2008-72507.

Iwata K, Moriwaki, T (1977), 'An application of acoustic emission measurement to in-process sensing of tool wear', *Ann CIRP*, **26**: 21–6.

Jemielniak K, Arrazola, P J (2008), 'Application of AE and cutting force signals in tool condition monitoring in micro-milling', *CIRP J Manufacturing Sci Technol*, **1**: 97–102.

Kanga I S, Kimb J S, Kangc M C, Leed K Y (2008), 'Tool condition and machined surface monitoring for micro-lens array fabrication in mechanical machining', *J. Mater Process Technol*, **201**: 585–9.

Kustas F M, Fehrehnbacher L L, Komanduri, R (1997), 'Nanocoatings on cutting tools for dry machining', *Ann CIRP*, **46**: 39–42.

Lan M S, Dornfeld D A (1982), 'Experimental studies of tool wear via acoustic emission analysis', *Proceedings of the 10th North American Manufacturing Research Conference*, Society of Manufacturing Engineers, Dearborn, MI, pp. 305–11.

Lu P, Chou Y K, Thompson R G (2009), 'AE single evolution in machining by diamond coated tools', *Proceedings of the ASME 2009 International Manufacturing Science and Engineering Conference*, American Society of Mechanical Engineers, New York, MSEC2009-84372.

Lu P, Chou, Y K, Thompson R G, (2010), 'Short-time Fourier transform method in AE signal analysis for diamond coating failure monitoring in machining applications', *Proceedings of the ASME 2010 International Manufacturing Science and Engineering Conference*, American Society of Mechanical Engineers, New York, MSEC2010-34305.

Moriwaki T, Tobito, M (1990), 'A new approach to automatic detection of life of coated tool based on acoustic emission measurement', *ASME J Engng Industry*, **112**: 212–18.

Mukhopadhyay C K, Venugopal S, Jayakumar T, Nagarajan R, Mannan S L, Raj, B (2006), 'Acoustic emission monitoring during turning of metal matrix composite and tool wear', *Materials Evaluation*, **64**: 323–30.

Somasundaram S, and Raman S (1993), 'Acoustic emission studies while machining with coated tools', *Trans NAMRI/SWE*, **XXI**: 83–94.

Sutoh T, Yabe H, Shinozaki N, Hiruma T, Sato Y, Nashida T, Kaneko, S (1997), "Gabor filter' technique for analyzing event-related desynchronization', *Electroencephalogr Clin Neurophysiol*, **103**: 154.

2

High-performance machining of austenitic stainless steels

Ana Isabel Fernández-Abia,[1] Joaquín Barreiro García[1] and Luis N. López de Lacalle[2]

Abstract: Turning has been widely studied and documented. However, limitations still remain due to the absence of reliable information about certain materials. Such is the case for austenitic stainless steels. Despite the fact that these materials are of great economic and technological significance, a lack of knowledge still exists about their behavior when machining. Nevertheless, stainless steels are one of the most widely used groups of materials, representing 10 per cent of the world's steel production, and austenitic steels comprise 70 per cent of all stainless steels. Characteristics that have a direct effect on the machining of these steels are analyzed here: the effects of the addition of different components on machinability, the effects that cutting parameters have over some thermomechanical properties by means of simulation tests, and the effects of cutting speed on tool wear, superficial quality, cutting forces and chip formation. These effects are studied for one of the most widely used austenitic stainless steels, AISI303, using both conventional (V_c < 350 m/min) and very high (up to 870 m/min) cutting speeds. Our final

[1] Department of Mechanical, Informatics and Aerospatiale Engineering, University of León, Campus de Vegazana, s/n. 24071, León, Spain.

[2] Department of Mechanical Engineering, University of the Basque Country, Alameda de Urquijo, s/n. 48013, Bilbao, Spain.

Published by Woodhead Publishing Limited, 2013

objective was to determine the behavior of austenitic steels under high-performance machining, with direct application to optimizing production processes.

Key words: Austenitic stainless steels, high-performance machining, high-speed machining, turning.

2.1 Introduction

The current highly competitive nature of industry has forced companies to increase the performance of their manufacturing processes. This provides a challenge for production companies to improve processes, so that product manufacturing can satisfy high manufacturing standards related to lead time, quality and cost. For the machining sector, new technologies have been developed in recent years, such as high-speed machining or high-performance machining. The main benefits of these technologies are the production of parts with improved superficial quality and higher dimensional precision, shorter manufacturing times and lower production costs. These improvements have been made possible through the evolution of machine tool technology, the development of new tool geometries and tool materials, and improved understanding of machined materials.

However, study of high-speed machining has focused in particular on milling operations in applications related to the mold industry, in either the aeronautical or the automobile sectors, production of stamps and dies, or machining of light alloys for the aerospace industry. On the other hand, although turning processes have been widely studied and detailed, limitations still exist for high-speed turning of particular materials, due to a lack of information about their behavior and characterisation. Such is the case of austenitic stainless steels; regardless of being materials of high economic and technological importance, there are still few reliable data regarding the machining of these materials under high-performance conditions.

Stainless steels represent 10 per cent of the world's steel production and austenitic steels comprise 70 per cent of these. In accordance with data from the International Stainless Steels Forum (ISSF), world production of stainless steel is experiencing continuous growth and it is expected that consumption will continue to increase each year. In 2010, an increase of 24.9 per cent took place, with production reaching a historical record of 30.6 million tons. In the first part of 2011, total production was 16.4 million tons. Figure 2.1 shows world production of stainless steel (in thousands of metric tons) over the last decade.

The reasons for the extensive use of stainless steels are obvious: stainless steel is a sustainable material with high durability. In addition, today environmental factors are of greater importance when selecting a material and stainless steels are unquestionably recyclable; it is possible to separate and recover them at the end of their useful life. Components

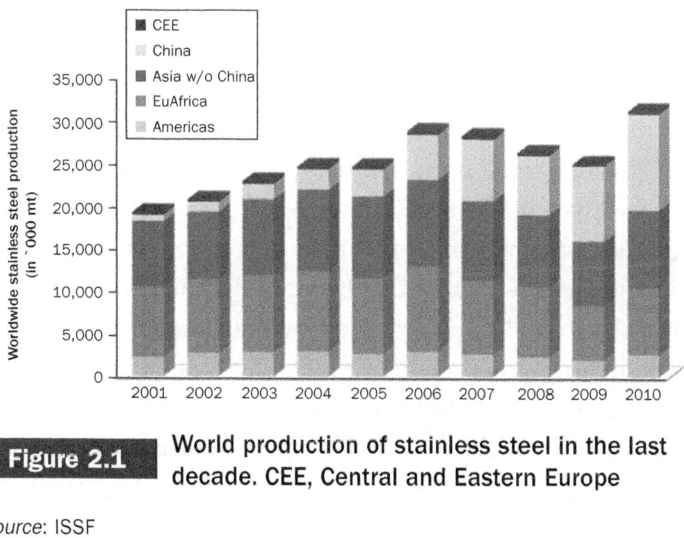

Figure 2.1 **World production of stainless steel in the last decade. CEE, Central and Eastern Europe**

Source: ISSF

of stainless steels such as chromium, nickel and molybdenum are valuable and make recycling economically feasible. In contrast to other engineering materials, stainless steel can be recycled without suffering degradation.

Among stainless steels, the most widely used are austenitic steels owing to their excellent mechanical properties and corrosion-resistance. Today, stainless steels are used in numerous sectors, and are essential in the chemical, food, medical, civil construction and automotive industries. A high percentage of products in these applications require machining operations.

One industry that needs stainless steels is the décolletage sector, where large series of parts of small and medium size are produced. In this sector the machining of components over very short production times is fundamental. Therefore, particular attention is given to turning operations of these types of materials to achieve improvement in performance, for example by using severe cutting conditions to reduce operation cycle time.

However, these materials are considered of poor machinability and tool makers recommend very conservative cutting speeds. The recommended range of cutting speeds for austenitic stainless steel turning is between 100 and 250 m/min. For this reason, research into austenitic stainless steel turning has been carried out in the scope of this conservative cutting speed range relating to economical tool life. However, when considering automatic or unmanned machine tools with a sliding head, multi-spindle and single-spindle CNC lathes or multitasking machines, where machine-hour cost is high, reducing operation times is essential for cost-effective operation. This objective is achieved by machining under severe cutting conditions. Although these conditions reduce tool life, in general the total cost charged over each machined part is reduced.

In most cases, an increase in cutting speed improves productivity and reduces significantly the cost of parts because it reduces machine-hour cost, both general and manpower. On the other hand, an increase in useful tool life has a marginal effect. Therefore, to reach high levels of productivity and to take advantage of the capacity of modern machine tools, it is necessary to explore and characterize the behavior of the material and cutting tool when working at severe cutting conditions, i.e. when using values for cutting parameters superior to those used traditionally.

In this chapter, several aspects related to the turning of austenitic stainless steels using severe cutting conditions are analyzed. We examine the effect of using very high cutting speeds on cutting forces, integrity of the machined surface, chip formation and tool wear.

2.2 Properties of austenitic stainless steels

Stainless steels are ferric alloys with a minimum 10.5 weight per cent chromium and a maximum 1.2 weight per cent carbon. Chromium is the main component providing resistance to corrosion, as it favors formation of a chromium oxide film that protects the underlying material from attack. Resistance to corrosion of these steels can be further improved with addition of other alloy elements such as nickel, molybdenum, nitrogen or titanium.

Remarkable changes are detected in the microstructure of stainless steels as a function of the content of carbon, chromium, nickel or other elements. Stainless steels can thus be classified in the following groups: ferritic, martensitic, austenitic, austenitic–ferritic (duplex) and hardened by precipitation.

Among stainless steels, austenitic grades are the most widely used due to their high resistance to corrosion, excellent mechanical properties, good weldability and low maintenance cost.

The typical alloy elements of austenitic steels are chromium (16–26 per cent) and nickel (6–22 per cent), with very low carbon content (<0.1 per cent), which can be increased up to 0.25 per cent when chromium and nickel contents are maximum. Nickel favors the gamma phase by enlarging the austenitic field.

Austenitic stainless steels have a single-phase structure formed by equiaxial austenite grains. The face-centered cubic (FCC) structure of austenite provides high ductility, good resistance to high temperatures and excellent resistance to impact, even at cryogenic temperatures. Austenitic stainless steels are not magnetic and their mechanical properties do not improve with thermal treatments. However, they can be hardened by cold-working.

Austenitic stainless steels have the largest number of alloys. Two groups may be mentioned here: chromium–nickel alloys (AISI300 series) and chromium–manganese–nickel alloys (AISI200 series). The first group is the most important and includes the most commonly used stainless steels. In the AISI300 series, the best-known representative is AISI304 steel, also called 18-8 steel (18 per cent Cr – 8 per cent Ni).

By modifying their chemical composition and adding particular alloy elements, a great variety of austenitic stainless steels are produced with different mechanical properties and resistance to corrosion. For example, addition of 2 per cent molybdenum transforms 304 stainless steel to 316 stainless steel, which offers higher resistance to pitting corrosion and improves behavior to corrosion under stress. When these materials are exposed to high temperatures (450–850 °C), chromium carbides are precipitated at grain borders, causing a chromium fault in adjacent regions. This phenomenon is known as sensitization. Areas poor in chromium are less resistant to corrosion and are susceptible to inter-grain corrosion when they come into contact with an acid environment. To avoid this, stabilizer elements such as titanium, niobium or tantalum are added, which increase sensitization temperature to 1250–1350 °C. Another solution is to reduce the carbon content to below 0.03 per cent to produce the so-called low-carbon content steels AISI304L, AISI316L and AISI317L. With these steels, sensitization only occurs when exposing them to very high temperatures for very long periods.

Figure 2.2 details different austenitic stainless steels and the effect of alloy elements.

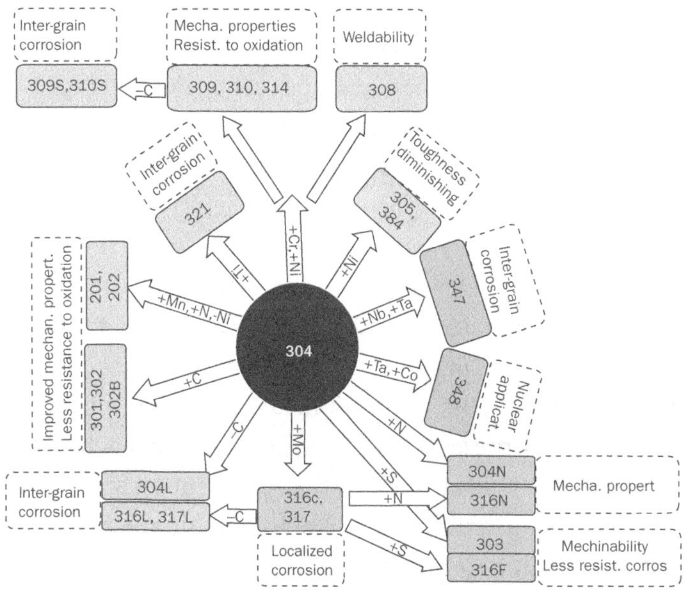

Figure 2.2 **Effect of alloy elements on AISI304 austenitic stainless steel**

Source: Di Caprio (1999)

2.2.1 Machinability of austenitic stainless steels

Austenitic stainless steels have numerous applications, especially in the food, drink and pharmaceutical industries. The wide use of these materials is due to their good hygienic properties and excellent long-term performance. They are used extensively in storage of liquefied gases, kitchen utensils, surgical instruments, transport, residual fluid treatment, civil construction, and the aeronautical and naval industries. They are widely used in the production of critical components in chemical and nuclear plants due to good mechanical properties and high resistance to corrosion. As result of the huge number of applications, machining of stainless steels has increased notably: most of these products require, at least, one machining operation during their production process. The importance of stainless steels machining is corroborated by data provided by Sandvik (a leader in the production of cutting tools): approximately 30 per cent of turning tools are produced for the machining of stainless steels.

In this context, it is evident that machinability properties are an essential feature for stainless steel manufacturers, as an improvement in machinability has a great impact on competitiveness. Development of stainless steels with improved machinability and improvement of machining processes for these materials leads to substantial reductions in the production cost of components, enlarging the range of applications for these materials. In general, the machinability of stainless steels is inferior to that of common steels, being even lower for austenitic types than for ferritic and martensitic types, mainly due to the following:

- The friction coefficient at the material–tool interface is, in general, higher for stainless steels than for other steels. This leads to more rapid degradation of tools. Among the different types of stainless steels, austenitic steels are characterised by higher friction coefficients.

- Stainless steels have a very low thermal conductivity compared with common steels, and is especially significant for the austenitic group. Heat generated in the cutting edge during machining operations cannot be dispersed easily through the machined part and accumulates in the cutting area. There is thus a temperature rise in the tool, which consequently suffers a decrease in life. Low thermal conductivity also affects formation of segmented chips.
- The coefficient of thermal dilation for austenitic stainless steels is higher than that for carbon steels. This means added difficulty in maintaining machining tolerances.
- Austenitic stainless steels present a large area of ductile strain. This explains the natural trend to produce very long chips or to generate a built-up edge. A common word for such steels is 'sticky'.
- The tendency to cold-work hardening is higher for stainless steels than for common steels. As a consequence, mechanical strength in the cutting area increases during chip removal, leading to higher forces over the cutting tool. Also, cold-work hardening implies an increase of superficial hardness in the machined part, which affects tool life in subsequent cylindering passes.

For all these reasons, austenitic stainless steels are considered to be difficult to machine. Their machinability improves slightly by adding alloy elements. For example, sulfur is added to generate globular inclusions of MnS in the main austenitic structure, which allows the use of higher cutting speeds and makes chip-breaking easier. This is the case of AISI303 stainless steel (0.3 per cent S). However, although improved machinability is achieved, addition of sulfur decreases mechanical properties and resistance to corrosion. An alternative that does not affect mechanical or corrosion properties is addition of elements such as calcium, copper or bismuth (Akasawa *et al.*, 2003; Liew, 2010).

Addition of inclusions to steel favors the formation of material layers that adhere to the tool surface, providing a positive lubricant action. This reduces the rate of tool wear considerably. There are different types of layers in this regard: layers coming from the work material, oxide layers due to chemical reactions between chip-tool and coolant, layers of plastic strain in the tool and material transfer layers (MTLs) or built-up layers (BULs).

In machining of austenitic stainless steels, both BUL and built-up edge (BUE) formation are of particular importance. A BUL is a layer that extends along the rake face of the cutting tool. This layer is usually of constant but lower thickness than the BUE that is formed in the tool tip.

Therefore, when sulfur-improved steels are being machined, a layer of MnS appears at the tool surface, which acts as a lubricant, reducing friction forces and reducing the temperature. The behavior of this layer in preventing tool wear is dissimilar at different cutting speeds. When machining at low or medium cutting speeds the MnS layer is stable and it acts as a protective barrier for tool wear. However, at high cutting speeds, the MnS layer becomes unstable, thin and viscous, and its capacity to prevent tool wear reduces significantly. As consequence, new steels have been developed with a controlled content of oxides (inclusions of CaO–MnO–SiO_2–Al_2O_3) obtained by de-oxidation of steel, with the objective of increasing the stability of the MnS protective layer at high cutting speeds. This treatment also improves machinability at conventional cutting speeds.

When cutting speed increases, a displacement of the MnS layer takes place from the tool rake surface toward the flank surface. In fact, what occurs is that this layer becomes unstable in the rake surface and it becomes more stable in the flank surface, so tool wear rate in the tool flank decreases and tool wear in the crater predominates.

Therefore, for machining of austenitic stainless steels at high cutting speeds it is convenient to have optimized cutting tools to support wear in the crater.

On the other hand, the low thermal conductivity of stainless steels prevents dissipation of heat generated in the cutting area, and therefore temperature increases. High temperatures together with high ductility of the austenite phase increase material adhesion on the tool surface, causing BUE formation. A BUE increases tool wear rate and can lead to uncontrolled tool edge breaking, resulting in superficial integrity of the machined part. Development of a BUE depends to a great extent on temperature and as such can be controlled via cutting speed. At high cutting speeds, temperature increases and strength of the BUE decreases, coming off the tool. The recurrent process of BUE formation and BUE detachment produces variations in cutting forces at low frequencies (100 Hz) causing irregular tool wear. This behavior makes it difficult to estimate tool life when machining stainless steels.

Work hardening with austenitic stainless steels produces hard surfaces and chips that may cause notch wear due to repetitive micro cracks. Also, austenite produces long and continuous chips that are difficult to break. Therefore, careful design of chip breaker geometry is necessary.

To face machining austenitic stainless steels under severe conditions, optimisation of tool geometry and tool material is essential. With regard to tool geometry, a very sharp edge is recommended to minimise material strain and produce a softer cutting with lower cutting forces. The use of physical vapor deposition (PVD) techniques for tool coating produces edges of high sharpness, whereas chemical vapor deposition (CVD) produces more rounded edges. The use of positive rake angles and wide chip breakers favors continuous cutting and smooth exit of chips, reducing the heat generated by friction.

The tool material must have good hardness at high temperatures, excellent resistance to abrasive and diffusion wear, and low tendency to adhesion. These properties are provided by coated carbide tools, which are the best option for machining austenitic stainless steels.

2.2.2 High-performance machining and high-speed machining

In the context of metal machining, improved competitiveness comes not only through improvements in productivity; strict dimensional precision and superficial quality are demanded at reduced time and cost. Today, the main objective of machining is to maximise material removal rate while minimising tool wear, satisfying GD&T (geometric dimensioning and tolerancing) and superficial quality, and carrying out unmanned operations ecologically.

This complex objective is termed 'high-performance machining', and in some sectors, such as automotive, aeronautics, machine tools or molds manufacturing, is considered to be strategic.

High-speed machining (HSM) allows the objectives of high-performance machining to be achieved. HSM involves a high feed rate and cutting speed, although there are no specific values for these parameters when a machining process is carried out at high speed. The term HSM is relative and depends on the machined material; significant differences in cutting speeds are found among different types of materials. For example, machining aluminum at 1.800 m/min is easier than titanium at 180 m/min due to the influence of tool life.

Considerable research in the field of HSM has been made in recent years, and has been related to diverse aspects: the mechanism of chip formation, evolution of cutting forces, cutting temperature, tool wear mechanism, and superficial quality or integrity of the part surface.

Several conclusions have been made regarding the effects of machining at high cutting speeds. The behavior of cutting forces in HSM cannot be predicted using Merchant's cutting model (Merchant, 1945) because this model does not consider cutting speed as an analysis factor. Other models exist, such as that developed by Mathew and Oxley (Hastings *et al.*, 1980), which do take into consideration the effect of cutting speed. According to the Mathew and Oxley model, cutting force decreases at moderate increases in cutting speed. Over a particular range of cutting speeds, depending on the specific behavior to work hardening of each material, cutting forces then increase progressively.

Various explanations have been given for the reduction of cutting forces with increase of cutting speed when working at low or moderate cutting speeds. The shearing angle increases with machining speed due to a reduction in thickness in the sliding area and friction coefficient at the chip–tool interface. An increase of shearing angle reduces the area of shearing plane and, consequently, cutting forces decrease (Sutter and Molinari, 2005). Other authors (Pawade *et al.*, 2007; Fang and Wu, 2009) have justified this behavior through material thermal softening at high cutting speeds. An increase of cutting speed causes an increase of temperature, thereby increasing deformation and flow of material, which in turn lowers cutting forces. On the other hand, the high cutting forces at very low cutting speeds can be explained by BUE formation.

At low and moderate cutting speeds, the effect of work hardening in some materials leads to an increase of cutting forces, while other materials experience adiabatic sliding that favors a reduction of cutting forces. This behavior has been explained by a change in shearing deformation when the strain rate exceeds a critical value. Heat generated by severe plastic deformation causes a local increase of temperature and therefore material strength decreases. On the other hand, due to the high strain rate the material suffers work hardening. When thermal softening overcomes the work hardening effect, a local strain band is formed through which chips can slip, leading to reduced cutting forces. The cutting speed at which this change in behavior is observed differs for different materials.

Other authors have focused on the mechanism of chip formation in HSM (Ekinovic *et al.*, 2002; He *et al.*, 2002; Dolinsek *et al.*, 2004). A transition from continuous to segmented chips is observed for most materials with increasing cutting speed. Sullivan *et al.* (1978) showed that the transition from continuous to segmented chips occurs at 35 m/min for austenitic stainless steels. This cutting speed is very low due to the poor thermal conductivity of austenitic stainless steels, the compact hexagonal structure and the high hardness of the segmented chip as a result of adiabatic slipping. The negative aspect of segmented chips is instability and chatter.

HSM improves superficial finish and therefore fewer finishing operations are required. Several researchers have shown that an increase in cutting speed produces a decrease in superficial roughness (R_a, average roughness) (Yousefi and Ichida, 2000; Davim, 2008). This due to the disappearance of BUEs at higher cutting speeds (Astakhov, 2006). And Chen (2000) explained this based on the lower plastic behavior of steels at higher strain rates.

However, the negative effect associated with HSM is the rapid degradation of the cutting tool. Temperature at the chip–tool interface increases at high cutting speeds, which accelerates tool wear significantly. The main type of wear at high cutting speeds is diffusion wear. The development of cutting tools that are able to resist high temperatures is therefore of primary importance. However, each operation/part material combination requires a specific tool material, as the mechanism of chip formation and tool wear changes for each case.

2.3 Machining study of austenitic stainless steels at high performance conditions

The results obtained from machining tests are discussed below. These results allow us to analyze the behavior of austenitic stainless steels in the context of high-performance machining.

2.3.1 Description of material and equipment

We focus on AISI303 steel, a chromium–nickel austenitic stainless steel modified by the addition of sulfur to improve machinability. This steel is widely used for the production of large series using unmanned machine-tools.

The chemical composition and mechanical properties of AISI303 steel are given in Table 2.1.

The most common shape for this steel is a rolled bar. When the bar needs to be calibrated, i.e. its dimensions better controlled, then it is stretched. This causes plastic deformation

Table 2.1 Chemical composition and mechanical properties of AISI 303

Chemical composition

	%C	%P	%S	%Si	%Mn	%Cr	%Ni	%Mo	%Ti	%N	%Cu	%Fe
AISI 303	0.050	0.033	0.273	0.365	1.776	17.773	8.783	0.271	0.003	0.041	0.273	70.392

Mechanical properties

	R_m (N/mm^2)	R_p 0.2% (N/mm^2)	R_p 1% (N/mm^2)	Z %	A % (%L_{sd})	Hardness HB
AISI 303	638	369	414	53	46	169

of inclusions on the outside of the bar, which makes them smaller. Figure 2.3 displays photomicrographs showing this effect and its influence on machining operations.

The structure of austenite is shown in these images together with the presence of manganese sulfurs. Sulfurs on the outside of the bar are smaller than at the bar center, although the percentage of these sulfurs is similar in both areas. The deformed area is less than 0.5 mm wide and hence it does not affect the cutting processes. Figure 2.4 shows a general scheme of the equipment and software used to carry out the experiments.

A dynamometric plate was used to measure cutting forces. This was installed in a tool holder so that forces are measured directly in the tool, as shown in Figure 2.5. The electric signal generated by a quartz transducer is amplified and filtered using a load amplifier that generates an output of ±10 V analog voltage. The conditioned signal is fed to a computer equipped with an acquisition card configured at 200 kHz sampling frequency. DasyLab software was used for handling

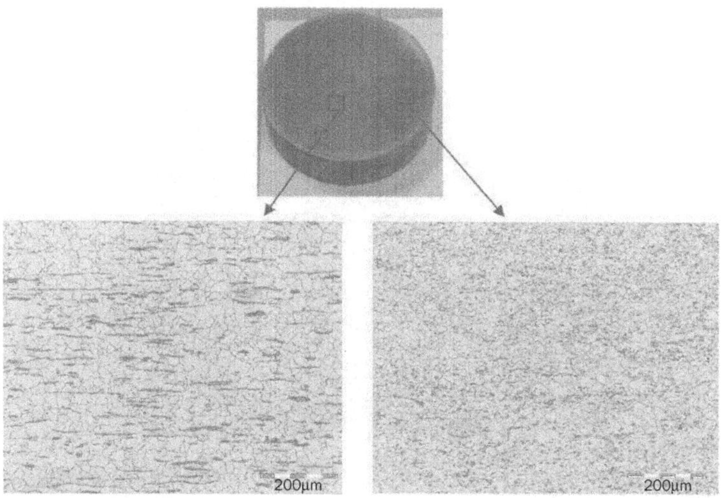

Figure 2.3 **Photomicrographs of AISI303 (×200)**

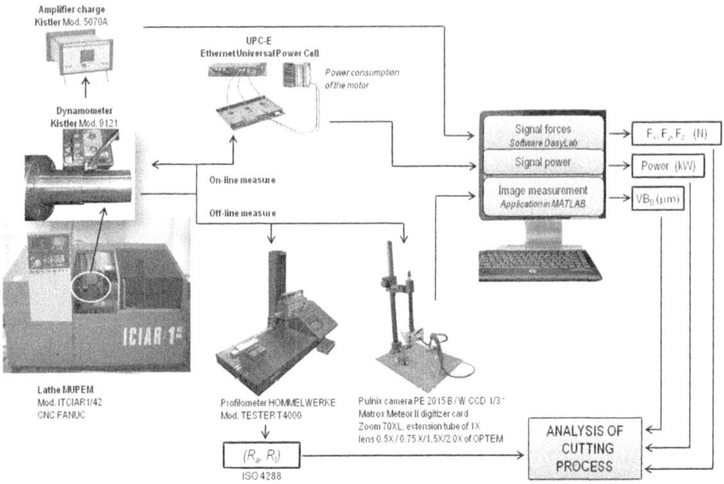

Figure 2.4 Scheme of the experimental equipment

Figure 2.5 Directions of the three components of cutting force acquired by a Kistler 9121 dynamometer

force signals. Average values were used for each of the three force components (F_x, F_y and F_z) along a machining pass.

Control of the superficial finish was performed with a Class 1 inductive perthometer with ±5 per cent accuracy. R_a (average roughness) and R_t (maximum depth of roughness) values were used as specified by the ISO4288 standard. Exploration length was 4.8 mm and cut-off 0.8 mm. Three measures were acquired over machined specimens and average values for each were calculated. Sampling areas for roughness evaluation are indicated in Figure 2.6; the ends of the machined surface were discarded to avoid the disturbing effect of tool input and output.

Images of the tool flank face, rank face and tool tip were acquired to measure tool wear. Tool wear was later measured based on the directives of the ISO3685 standard. The equipment used for image acquisition included a 1/3-inch CCD B/W camera with a frame grabber card, an optical system comprising an OPTEM 70XL industrial zoom with a 1× extension tube and 2.0× lens. A regulated light source provides an intense cold illumination (Figure 2.7).

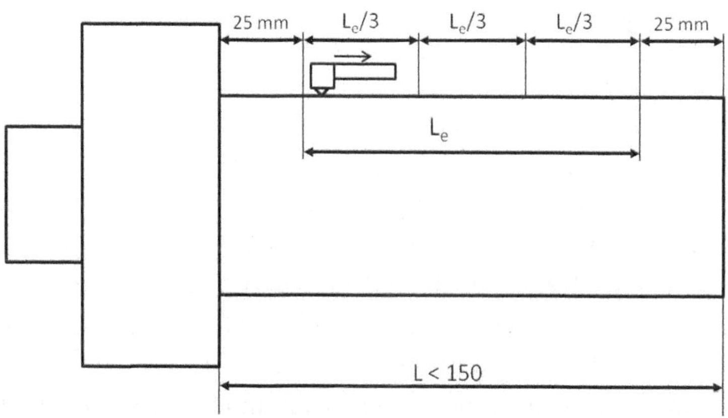

Figure 2.6 Areas for roughness measurement in the machined specimen

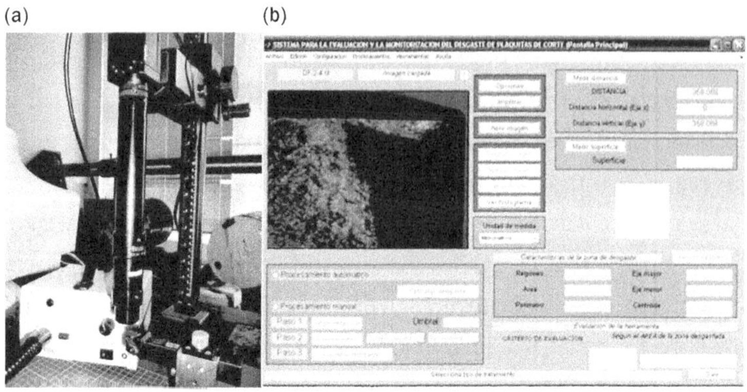

Figure 2.7 (a) Equipment for image acquisition, and (b) software developed for tool wear measurement

Additionally, scanning electron microscopy (SEM) of 3.5-nm resolution and energy-dispersive X-ray spectroscopy (EDX) were used.

2.3.2 Definition of the optimal parameter window for high-performance turning of austenitic stainless steels

Simulation of the cutting process allowed us to carry out virtual design of the experiments, from which information could be extracted regarding viability of the process. The influence of a particular geometry or specific tool material on cutting forces and temperature could thus be analyzed. Finite element analysis is widely used for calculation of stress and thermal fields and for analyzing shearing areas (primary, secondary and tertiary) that appear in the material when it is machined with different tool geometries. Study of tool heating during machining is also of great interest, as it has important consequences for determining tool life.

In the context of high-performance machining, a series of simulations were carried out with the objective of analyzing the influence of various process parameters (cutting speed, feed rate, depth of cut) on cutting forces, temperature at the chip–tool interface, generated stresses and chip geometry.

Simulations were carried out with AdvantEdge™ software developed by ThirdWave Systems specifically for analysis of machining operations. This software integrates finite element numerical techniques and material modeling using Lagrange formulation with remeshing. Maranhao & Davim (2010) used this software to check that the predicted results are close to those obtained experimentally when machining austenitic stainless steel.

The following stages were performed for the machining tests:

Stage 1. First tests were carried out to validate the AISI303 steel constitutive equation included with the software. Validation tests were used to compare the tangential component of the cutting force, obtained experimentally with the dynamometric plate, with that obtained by simulation using AdvantEdge. The turning tests covered a wide range of cutting conditions (Table 2.2), including cutting speeds typical of high-performance machining for austenitic stainless steels. Simulations were performed with the orthogonal cutting module described in Figure 2.8 where part and tool are represented.

To model the thermo-viscous-plastic behavior of the work material, the software uses a constitutive equation that is a variant of the Johnson–Cook law considering work hardening, strain rate and thermal softening, according to:

Table 2.2 Cutting parameters used in validation tests

Test no.	V_c (m/min)	f_n (mm/rev)	a_p (mm)
1	450	0.1	1
2	450	0.2	1
3	450	0.1	2
4	450	0.2	2
5	600	0.15	1.5
6	750	0.1	1
7	750	0.2	1
8	750	0.1	2
9	750	0.2	2

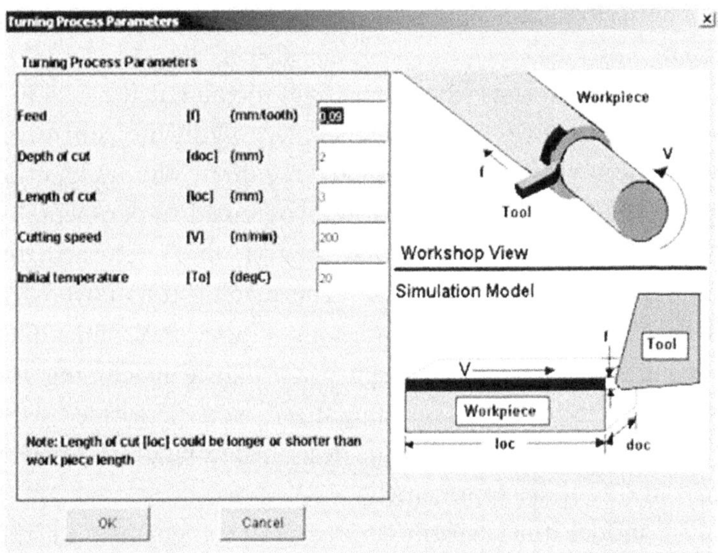

Figure 2.8 Orthogonal cutting scheme

Source: Third Wave Systems

$$\left(1+\frac{\varepsilon_p}{\varepsilon_0^p}\right)=\left(\frac{\bar{\sigma}}{g(\varepsilon_p)}\right)^{m_1} \quad si \;\; \varepsilon_p \le \varepsilon_t^p$$

$$\left(1+\frac{\varepsilon_p}{\varepsilon_0^p}\right)\left(1+\frac{\varepsilon_t}{\varepsilon_0^p}\right)^{m_2/m_1}=\left(\frac{\bar{\sigma}}{g(\varepsilon_p)}\right)^{m_2} \quad si \;\; \varepsilon_p > \varepsilon_t^p \qquad [2.1]$$

$$g=\sigma_0 \cdot \varphi(T) \cdot \left(1+\frac{\varepsilon_p}{\varepsilon_0^p}\right)^{1/n}$$

where:

- $\bar{\sigma}$ is effective stress (Von Mises stress)
- g is yield stress
- ε_p is plastic strain
- ε_0^p is strain rate of reference
- m_1 and m_2 are sensibility exponents to low and high strain rate, respectively
- ε_t is strain rate limit between the two regimes
- n is an exponent of work hardening
- T is temperature
- σ_0 is initial yield stress at the reference temperature T_0
- $\varphi(T)$ is a factor of thermal softening in the range 0–1.

Input parameters to perform the simulation are summarised in Table 2.3.

Figure 2.9 shows the results obtained. Experimental forces are higher than those obtained by means of simulation in tests carried out at a feed rate (f_n) of 0.1 mm/rev and they are lower at f_n = 0.2 mm/rev. The figure also indicates that the trend in force evolution is similar both in experimental and in simulated tests, with average deviations of about 10 per cent.

Table 2.3 **Input parameters for simulation with ThirdWave AdvantEgde**

Workpiece	
Workpiece length	5.0 mm
Workpiece height	2.0 mm
Workpiece material	AISI 304
Hardness	169 HB
Tool	
Cutting edge radius	0.02 mm
Rake angle	–6°
Relief angle	6°
Tool material	Carbide, Grade M
Coating material (three layers)	TiC: 0.03 mm Al_2O_3: 0.02 mm TiN: 0.02 mm
Process	
Feed	0.1–0.15–0.2 mm/rev
Depth of cut	1–1.5–2 mm
Length of cut	5.4 mm
Cutting speed	450–600–750 m/min
Initial temperature	20 °C
Friction	Default
Coolant	Not used
Simulation	
Standard mode	
Chip breakage	

Once the model has been validated, it can be used to obtain a first approach to the effects that cutting parameters have on specific thermo-mechanical variables, such as temperature, stress and strain. These aspects provide wide knowledge of the processes and constitute the starting point for optimisation of cutting operations, helping in selection of cutting conditions and tool geometries.

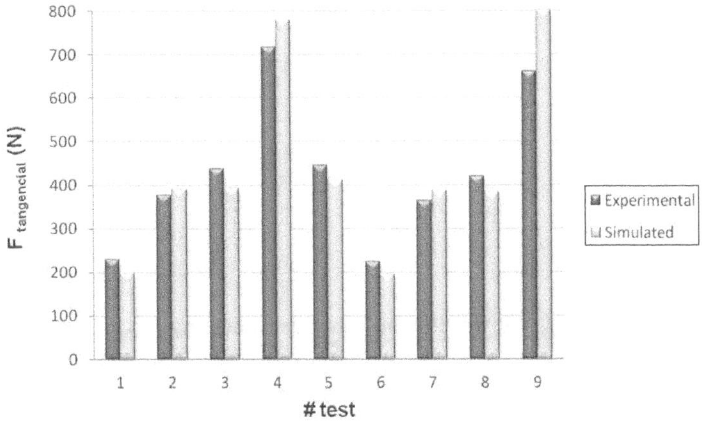

Figure 2.9 **Comparison between simulated and experimental tangential forces over a wide range of cutting conditions**

Stage 2. The next task was to carry out a series of simulations to analyze the influence of cutting speed on cutting forces, shearing angle, temperature at the chip–tool interface, strain rate and chip morphology. A wide range of cutting speeds between 37 and 870 m/min were used with feed rates of 0.1 and 0.2 mm/rev. Figure 2.10(a) shows the cutting forces obtained. There is a small decrease of the forces at higher cutting speeds, and this is more pronounced at higher feed rates (equivalent to higher chip load).

Figure 2.10(b) shows normalised cutting forces with regard to chip load. The specific cutting force is higher for smaller chip loads, i.e. material removal is energetically more expensive when chip load diminishes.

When shearing angle is measured for the two feed rates (or chip loads), an increase is observed with cutting speed (Figure 2.11). This increase in shearing angle leads to an increase of the cutting relationship (h/h_c) and, as result, cutting forces and strain decrease.

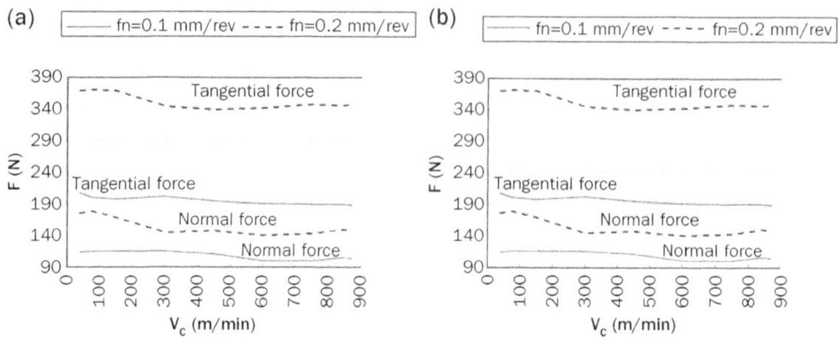

Figure 2.10 **(a) Relationship between cutting force and cutting speed, and (b) normalised cutting forces**

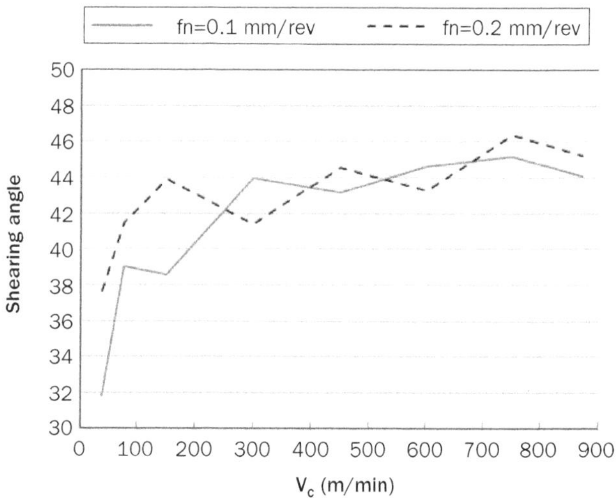

Figure 2.11 **Relationship between shearing angle and cutting speed**

Figure 2.12 plots temperatures in the tool rake and flank surfaces for different cutting speeds. The zero point on the horizontal axis corresponds to the rake face and flank face intersection. The highest temperature in the tool area is located at a certain distance along the edge, in the tool rake surface, after which there is a sudden decrease of temperature. This decrease is more pronounced for higher cutting speeds.

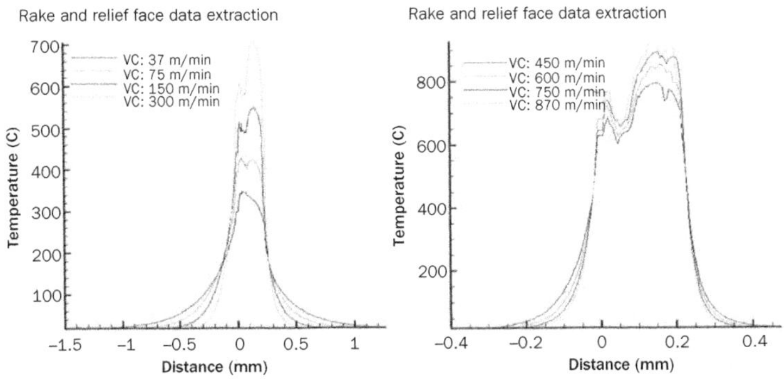

Figure 2.12 **Effect of cutting speed on tool temperature**

The ability to predict cutting temperature and its distribution is important in determining the cutting speed that maximises material removal rate while controlling tool wear. The chemical components of the tool can be affected by a diffusion process if a critical temperature is reached. Figure 2.12 reveals a substantial increase of temperature at higher cutting speeds, reaching values above 800 °C for cutting speeds higher than 450 m/min. Given that tool carbides do not bear temperatures above 800 °C, machining at cutting speeds over 450 m/min is not recommended when using uncoated carbide tools. However, when using coated carbide tools the coating (aluminum oxide layer) works as a thermal barrier, and therefore machining at high temperatures is possible.

The thermal energy generated per unit of volume and temperature in the chip–tool interface increases at higher cutting speed (Figures 2.13 and 2.14). This increase is greater above 450 m/min. In machining operations, a great amount of energy is transformed to heat associated with plastic deformation. If strain rate is high, or material thermal conductivity is low, then there is not enough time to dissipate the heat to the material surroundings and therefore the tool

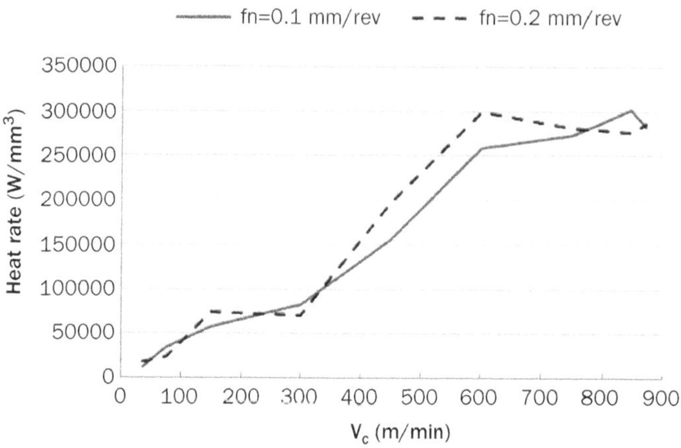

Figure 2.13 Thermal energy by unit of volume with respect to V_c

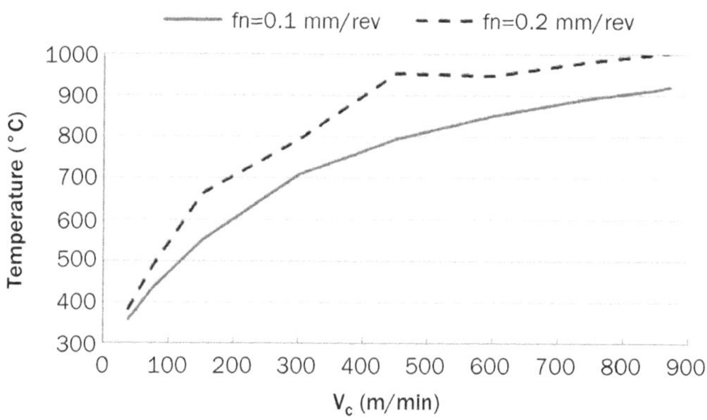

Figure 2.14 Relationship between temperature in the tool-chip interface and V_c

heats up. This temperature increase causes material thermal softening, inducing a decrease in cutting forces and in the energy required for deformation process.

Figure 2.15 shows material strain rate for different cutting speeds. Strain rate is higher at greater cutting speeds, as the

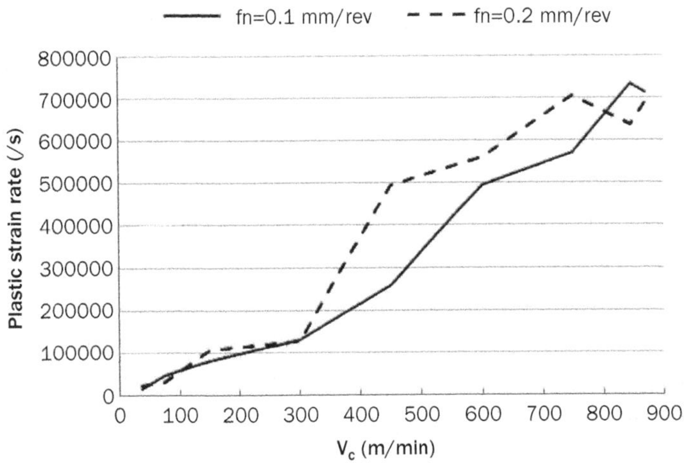

Figure 2.15 **Plastic strain rate versus cutting speed (V_c)**

tool moves more quickly and therefore material deformation is faster.

During cutting the strain rate reaches values between 10^4 and 70×10^4 s^{-1} for the highest cutting speeds. The behavior of steels at high strain rates is different from that under static or quasi-static conditions. Several authors have studied the effect of strain rate on plastic deformation of stainless steels and their micro-structural evolution (Woei-Shyan and Chi-Feng, 2001; Umbrello *et al.*, 2007). These studies have shown that plastic deformation and grade of work hardening are sensitive to temperature and strain rate. Therefore, to model the behavior of materials under these conditions, equations that consider the effects of strain rate and temperature on material plastic deformation are necessary. These equations must be valid over a wide range of cutting speeds and temperatures. The equations developed by Zerilli-Armstrong and Johnson-Cook, among others, are commonly used to simulate such processes.

Values for chip thickness were measured (Figures 2.16 and 2.17). Chip thickness was lower at increased cutting speeds. Lower chip thickness facilitates chip deformation, increases bending radius and therefore makes chip breaking easier. High cutting speeds are thus beneficial for chip breaking.

Stage 3. Further simulations were carried out with the objective of determining specific cutting energy and size effect in turning of AISI303 steel.

A parameter indicating process efficiency is the energy consumed by a unit of volume of removed material, i.e. specific cutting energy (P_s):

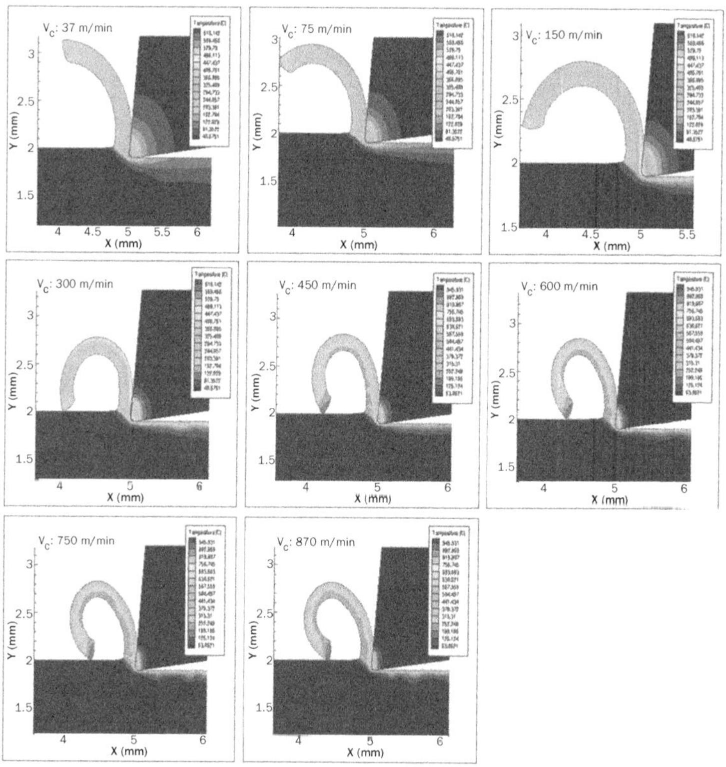

Figure 2.16 Chip formation versus cutting speed (f_n = 0.1 mm/rev)

$$P_s = \frac{F_c}{A_c} \qquad [2.2]$$

where F_c is the main component of the cutting force and A_c is the section of non-deformed chip. The value of the specific cutting energy is useful to provide a first idea of the forces required for machining a specific material. This parameter varies considerably for each material and is affected by changes in cutting speed, feed rate and tool geometry. However, in a first approach, for a particular tool rake angle and high cutting speed and feed rate, values of P_s can be considered to be constant. The value of P_s obtained in our simulation tests was around 1700 N/mm^2.

However, above a threshold value of non-deformed chip thickness, P_s experiences exponential growth. When using small chip sections, specific cutting force and power are higher. This behavior is known as a 'size effect' and refers to the increase of specific cutting energy for low values of non-deformed chip thickness. This phenomenon results from the fact that tool edge radius can no longer be considered to be

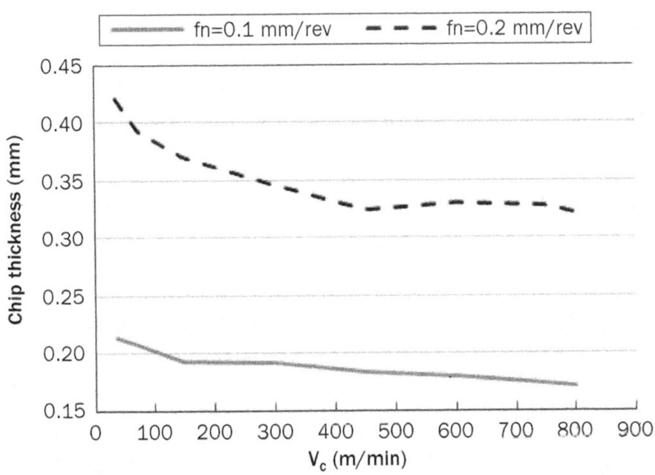

Figure 2.17 **Chip thickness versus cutting speed**

insignificant. In this case, the distribution of cutting force along the tool edge radius needs to be considered. The resulting force over the tool comprises the force in the cutting speed direction and thrust force. Thrust force does not contribute to material removal; it is the combination of the force that acts on the tool tip and the friction force acting on the tool flank surface. Thrust force remains constant at decreasing cutting thickness, so that for high values of non-deformed chip thickness thrust force constitutes a small proportion of total force. However, for small values of non-deformed chip thickness thrust force is proportionally high and cannot be ignored.

This effect was simulated using ThirdWave software obtaining P_s curves for different chip thickness. A set of simulation tests were carried out using the cutting conditions indicated in Table 2.4. P_s curves were obtained for different cutting depths while maintaining cutting speed at 450 m/min. The results are shown in Figure 2.18. Specific cutting energy grows exponentially for values of feed rate lower than 0.1 mm/rev.

Figure 2.19 shows the evolution of cutting force components with feed rate. A linear dependency is observed.

Simulations of the cutting operation by finite element analysis provide information about chip removal without the need for expensive experimental tests. It proved useful to analyze changes in cutting forces, cutting temperature, shearing angle, strain rate, chip morphology and chip breaking

Table 2.4 Cutting conditions used in the simulation tests with ThirdWave AdvantEdge

V_c (m/min)	450
a_p (mm)	1–1.5–2–2.5
f_n (mm/rev)	0.03–0.05–0.07–0.1–0.2–0.3–0.4

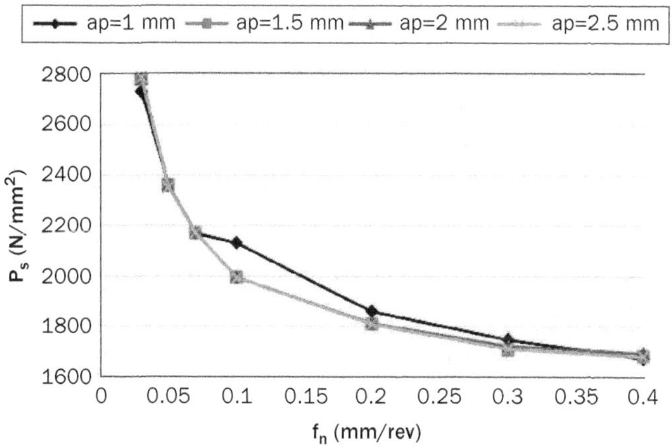

Figure 2.18 Effect of non-deformed chip thickness on specific cutting energy (P_s)

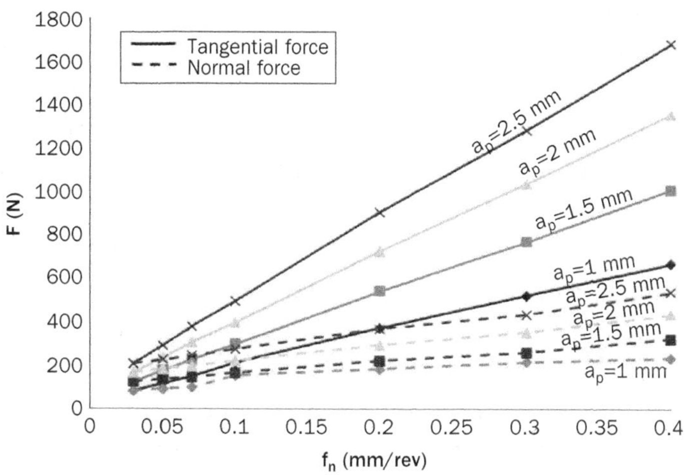

Figure 2.19 Relationship between cutting forces and feed rate

for several cutting conditions. These variables show considerable variations mainly due to changes in cutting speed.

The results provide a first approach to adjust cutting conditions and provide data about several parameters that are difficult to measure. However, experimental tests are required to complete and adjust the simulation results. Finite element analysis makes use of simplifications and suppositions ignored by the user. These simplifications can lead to variations in the results and consequently model validation is necessary. The next section affords this validation for AISI303 stainless steel.

2.3.3 Effect of cutting speed in high-speed machining of austenitic stainless steels: experimental study

This section describes the results obtained from machining tests using very high cutting speeds. Machining was carried out using AISI303 austenitic stainless steel. Machining operations were longitudinal turning and part specimens were bars of 180 mm length and 60 mm diameter. Coolant was not used in order to analyze the behavior of stainless steels under dry machining conditions. Dry machining satisfies the current trend of ecological machining, and does not modify the results when making comparative analysis of the chemical composition of worn tool surfaces.

Machining tests were carried out over a wide range of cutting speeds (Table 2.5), beyond the range recommended by tool manufacturers (180–250 m/min). Feed rate and depth of cut were fixed at 0.2 mm/rev and 1 mm, respectively, which are within the range of cutting conditions recommended by tool manufacturers. As discussed, they were not modified due to the adverse effect that these parameters have on the machined surface quality and cutting forces during turning.

Table 2.5 Cutting conditions used in turning tests

f_n (mm/rev)	0.2								
a_p (mm)	1								
V_c (m/min)	37	75	150	300	450	600	750	845	870

A 5.5-µm CVD multi-layer coated (TiCN, Al_2O_3, TiN) cemented carbide cutting tool was used, with sub-micrograin size and 8–10 per cent cobalt binder. The tool was a commercial-grade insert with TNMG 160408 geometry.

All machining tests were carried out using new edges without wear. Cutting forces, part roughness and superficial integrity, tool wear and chip geometry were measured during the operation. Nine metallographic probes, one for each cutting speed, were used to analyze superficial damage suffered by part material. Figure 2.20 shows one of the probes with corresponding metallographic image and marks corresponding to micro-hardness at different depths.

2.3.3.1 Influence of cutting speed on cutting forces

The evolution of tangential cutting force (F_t) as a function of cutting speed is given in Figure 2.21. The tangential cutting

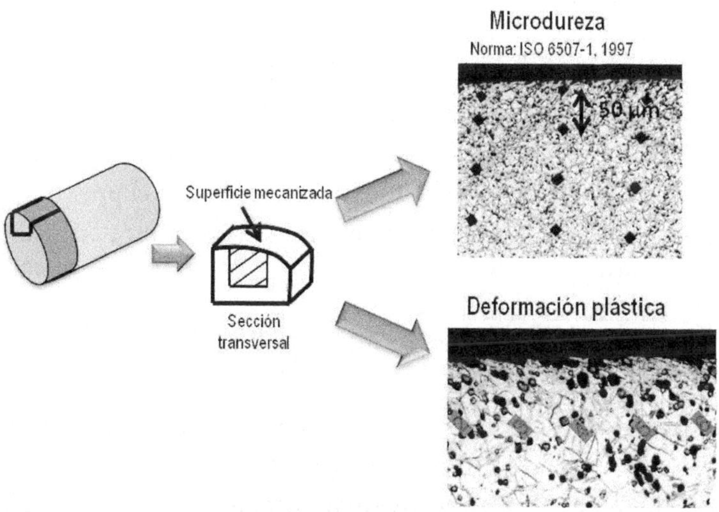

Figure 2.20 Probe, photomicrograph and micro-hardness measurement for AISI303

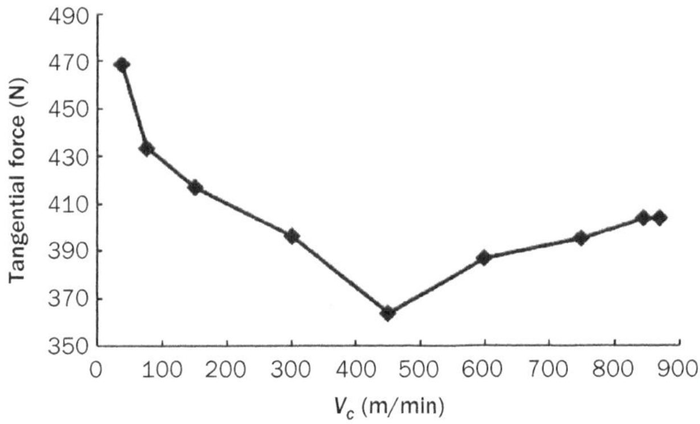

Figure 2.21 Relationship between tangential cutting force and cutting speed (f_n = 0.2 mm/rev; a_p = 1 mm)

force component is approximately twice the other two force components. Note that when cutting speed is increased from 35 to 450 m/min, cutting force first decreases and then increases. This behavior has also been observed in machining of titanium alloys (Sun *et al.*, 2009) and medium carbon steels (42CrMo4) (Sutter and Molinari, 2005).

This change in the trend of cutting forces with cutting speed can be explained by the balance between strain rate hardening and thermal softening, which are found in the well-known Johnson–Cook equation (Eq. 3.4) explaining the viscoplastic behavior of materials. This law expresses the flow stress (σ) as a function of strain (ε), strain rate ($\dot{\varepsilon}$) and temperature (T):

$$\theta = [A + B \cdot \varepsilon_p^N]\left[1 + C \cdot \ln\left(\frac{\dot{\varepsilon}_p}{\dot{\varepsilon}_0}\right)\right][1 - T_H^M] \qquad [2.3]$$

In the first part of the curve in Figure 2.21 (below 400–450 m/min) cutting forces decrease with increasing cutting speed. In this case, the thermal softening factor $[1 - T_H^M]$ predominates in the Johnson–Cook equation. Due to the low thermal conductivity of AISI303 steel (15 W/mK), heat generated during machining cannot be dissipated quickly. This causes an important increase in temperature in the chip deformation areas and consequently thermal softening of the work material; there is therefore a decrease in the cutting forces.

In the second part of the curve, cutting forces increase with further cutting speed increases. Now, strain rate hardening factor $\left[1 + C \cdot \ln \frac{\dot{\varepsilon}_p}{\dot{\varepsilon}_0}\right]$ predominates in Eq. 2.3. At these cutting speeds strain rate is high, around $10^5\ s^{-1}$; the significant sensibility of austenitic stainless steels to strain rate leads this factor to predominate over thermal softening and, as a consequence, cutting forces tend to increase. This agrees with the classic theory exposed by Oxley that considers the effect of work hardening, strain rate and temperature on the properties of materials (Oxley, 1989).

Therefore, 450 m/min is the cutting speed at which a change in behavior of the material takes place during the shearing process.

2.3.3.2 Influence of cutting speed on superficial finish

Table 2.6 shows the values obtained for surface roughness and maximum peak to valley height in the machined surface for different cutting speeds. These values are calculated as the average of six measurements.

Theoretical values of R_a and R_t can be estimated by means of the well-known Eqs 2.4 and 2.5, respectively. These equations consider roughness as a function of feed rate (f_n) and tool nose radius (r_ε). Accordingly, roughness values for the cutting conditions used in the tests are 1.5 and 6.25 μm, respectively. These theoretical values are lower than the

Table 2.6 Arithmetic average surface roughness (R_a) and maximum peak to valley height (R_t) for different cutting speeds (bold: minimum values: underlined: maximum values).

V_c (m/min)	37	75	150	300	450	600	750	845	870
R_a (μm)	<u>2.21</u>	1.84	2.00	2.14	2.12	**1.66**	<u>2.17</u>	1.84	1.90
$R_{a\,max} - R_{a\,min}$	1.50	0.58	1.11	0.74	0.26	0.16	0.11	0.15	1.05
R_t (μm)	<u>19.56</u>	9.50	10.77	10.47	10.91	**8.84**	<u>13.02</u>	10.45	12.61
$R_{t\,max} - R_{t\,min}$	17.62	3.28	8.36	5.96	5.62	1.49	2.26	2.93	10.27

measured values, due to surface finish being conditioned by additional factors related to machining stability, work material, tool wear or machining conditions.

$$R_a = \frac{f_n^2}{32 \times r_n} \quad [2.4]$$

$$R_t = \frac{f_n^2}{8 \times r_n} \quad [2.5]$$

Among the experiments, the surface machined at 600 m/min presents the lowest R_a (1.66 μm) and R_t (8.84 μm) values. Surfaces machined at 37 and 750 m/min are characterized by higher R_a and R_t values due to vibrations (unstable machining) during cutting. The variation in amplitude of cutting forces ($F_{max} - F_{min}$) is maximum for the three force components at 37 and 750 m/min, giving an idea of the instability of the cutting process. Meanwhile, the minimum force amplitude is reached at 600 m/min (Figure 2.22), which indicates stable

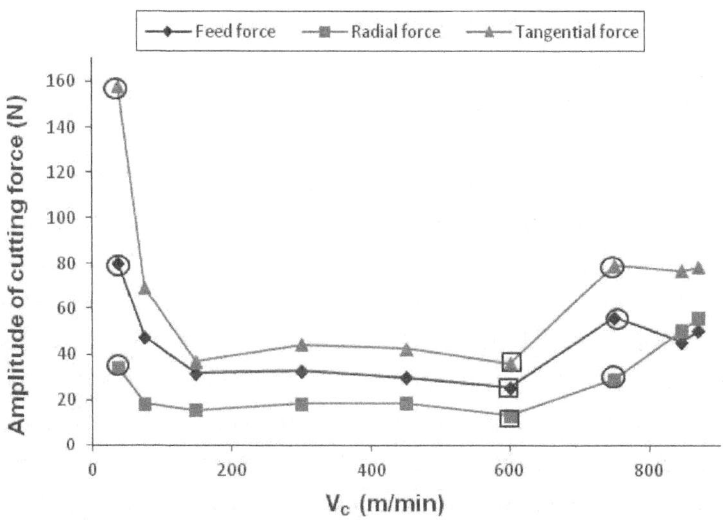

Figure 2.22 Relationship between cutting force amplitude and cutting speed

machining conditions and minimum values of roughness, as previously indicated. This corroborates the idea that roughness depends on cutting parameters and tool vibrations (Abouelatta and Mádl, 2001; Lu *et al.*, 2010).

2.3.3.3 Influence of cutting speed on part surface integrity

The action of machining modifies the part surface. The affected layer can reduce the mechanical properties and resistance to corrosion of austenitic stainless steels, reducing the useful life of products. Fatigue, fluency and corrosion initiate in the surface layer and they depend, to a great extent, on surface quality. Therefore, analysis of the condition of the machined part surface is necessary.

Scanning electron micrographs of the part surface were acquired to evaluate superficial defects caused by the turning operation at different cutting speeds (Figure 2.23). The change in the behavior of the material discussed above is also detected in the quality of machined surfaces at different cutting speeds. Surfaces machined at low cutting speeds (37–300 m/min) show cavities, metal debris and feed marks with smeared material particles, in part due to BUE appearance.

By contrast, surfaces machined at high cutting speeds (150–870 m/min) do not contain cavities or metal debris, but show evidence of material side flow. Similar effects have been reported during machining of modified AISI420 stainless steel (Liew *et al.*, 2003). Material side flow is defined as the displacement of portions of work-piece material in an opposite direction to tool feed movement. Material side flow produces burrs in the crests of feed marks. Material side flow takes place when chip material near the tool edge is exposed to high pressure and temperature. This produces complete plasticisation of the material, which flows through the main cutting edge toward the secondary cutting edge and adheres to the machined surface.

Kishawy and Elbestawi (1999) identified two mechanisms relating to material side flow. In the first, material is squeezed between the tool flank face and the machined surface when chip thickness is less than a minimum value. In the second mechanism, high temperatures during cutting lead to plasticisation of the material in the cutting zone, which flows through the worn edge toward the secondary edge. Both mechanisms can coexist simultaneously.

With regard to the first mechanism proposed by Kishawy, in our study chip thickness was measured at different cutting speeds in order to analyze the influence of chip thickness on the formation of material side flow. Figure 2.23 shows that

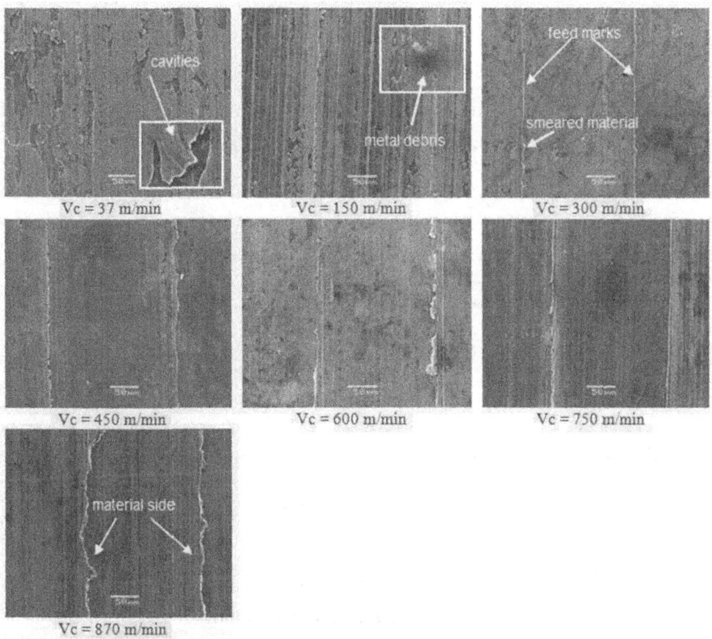

Figure 2.23 Scanning electron micrographs of the machined surfaces at different cutting speeds

evidence of material side flow is apparent at cutting speeds of 450 m/min, exactly when chip thickness is reduced (Figure 2.24).

The second mechanism is favored by increasing cutting speed and tool wear. Several authors (Warnecke and Bach, 1988; El-Wardany *et al.*, 1993) have demonstrated that cutting speed has a significant influence on this effect, as an increase of cutting speed increases cutting temperature, thereby favoring plasticisation of the material. On the other hand, tool wear in the secondary edge increases displacement of material from the shearing area toward the worn tool area. Wearing in the secondary edge was observed during machining tests when cutting speed was above 450 m/min.

In cutting operations part material is subjected to severe thermal effects (high temperatures and fast cooling) and mechanical effects (high stresses and strains) that are largely responsible for structural and hardness alterations in the part material. A study was performed to analyze the effect of cutting speed on part material properties.

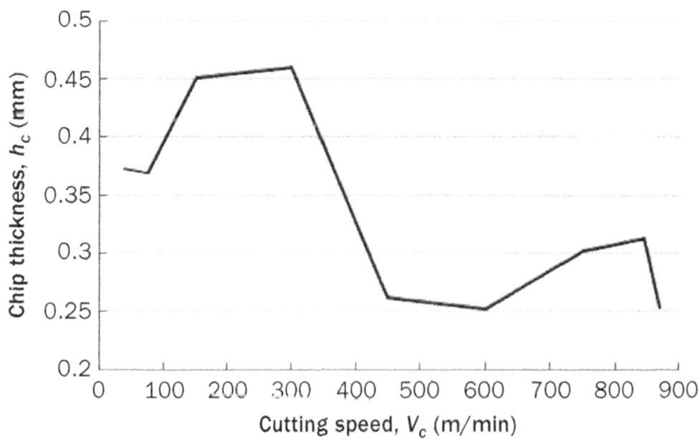

Figure 2.24 Relationship between chip thickness and cutting speed

Figure 2.25 shows micro-hardness measured at different depths under a machined surface for different cutting speeds. Micro-hardness was measured following the procedure stated for Vickers hardness measurement (ISO6507-1, 1997). Hardness measures were taken every 50 μm below the surface to 700 μm depth. A final measure was carried out at 1 mm depth. The average value of three measures was reported.

Material hardness is modified up to 600 μm below the machined surface. The trend was the same for all tested cutting speeds: the highest values are reached next to the machined surface and tend to diminish gradually toward the interior of the material where the effects of temperature and deformation are less severe.

Similarly, surfaces machined at high cutting speed had higher hardness values, which indicate the prevalence of work hardening over thermal softening, as indicated previously. Work hardening of the superficial layer during cutting can affect later passes. Therefore, cutting depth should be higher than the hardened layer, which is around 600 μm for the conditions used here.

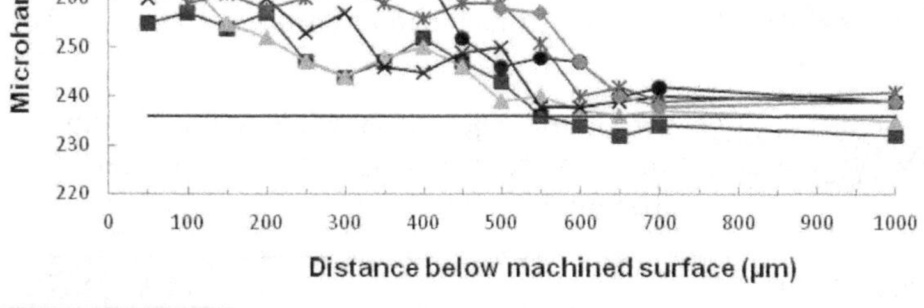

Figure 2.25 Average microhardness values

Figure 2.26 provides images of the superficial layer. Grain deformation is observed due to high pressure in the material during cutting. The width of the deformed area was estimated based on ten measurements in a transverse section. An analysis was performed for different cutting speeds (Table 2.7). The results show an increase in the depth of deformation with increasing cutting speed, due to work hardening suffered by austenitic stainless steel.

2.3.3.4 Influence of cutting speed on tool wear

The surface of cutting tools used in the machining tests was analyzed using SEM and EDX techniques to determine the nature of wear. As indicated, cutting tools were multilayer-coated, cemented carbide inserts ($TiCN–Al_2O_3–TiN$).

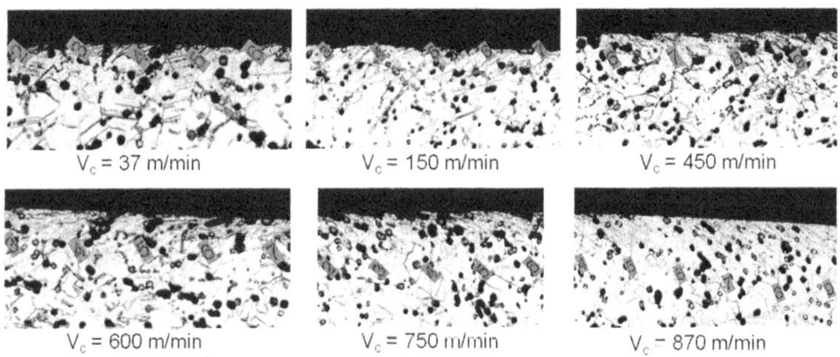

Figure 2.26 Depth of deformed microstructure for specimens machined at different cutting speeds

Table 2.7 Work hardening measures carried out in the transverse section of specimens machined with different cutting speeds

V_c (m/min)	37	150	450	600	750	870
Deformed width (μm)	2.18	1.84	3.16	3.57	5.11	6.05

Machining of austenitic stainless steels is characterized by formation of a coating layer in the tool edge that is rich in oxides and sulfurs originating from the steel. Figure 2.27 shows material layers adhering to the tool flank face. Quantitative element mapping via EDX analysis was used to identify specific components attached to the flank surface (Figure 2.28). This revealed that these layers are composed of iron (Fe), chromium (Cr), nickel (Ni) and manganese (Mn). As aforementioned, these elements come from AISI303 austenitic stainless steel. When machining in the range of moderate cutting speeds (37–300 m/min) these layers are located close to the main cutting edge and are spread over the flank face.

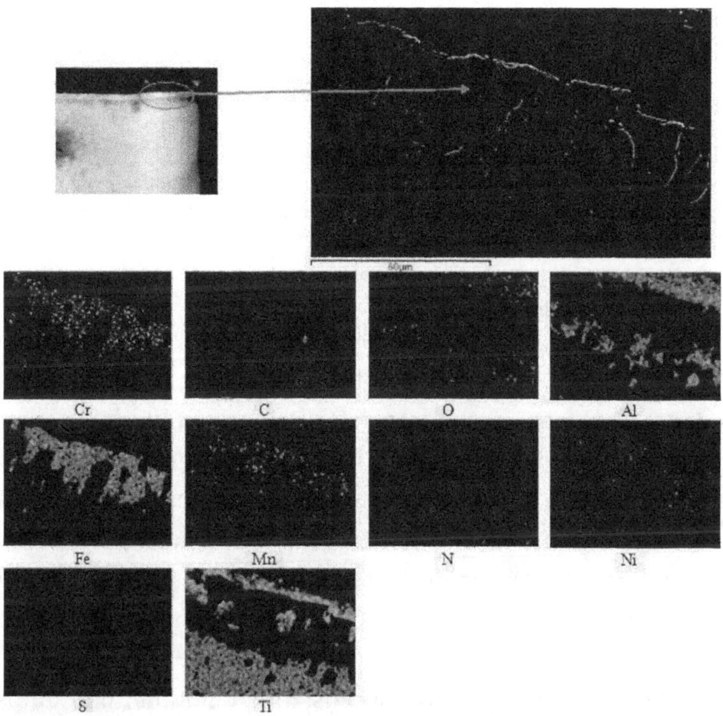

Figure 2.27 SEM mapping images of adhered material in the tool flank face

	New tool		Worn tool	
Element	Weight%	Atomic%	Weight%	Atomic%
C K	7.50	12.37	5.51	11.00
N K	31.82	44.97	10.82	18.53
O K	18.19	22.51	24.21	36.31
Al K	8.15	5.98	14.70	13.07
S K	-	-	0.21	0.16
Ti K	34.25	14.15	23.83	11.94
Cr K	-	-	4.15	1.91
W	0.09	0.01	-	-
Mn K			0.56	0.24
Fe K			14.39	6.18
Ni K			1.64	0.67
Totals	100.00		100.00	

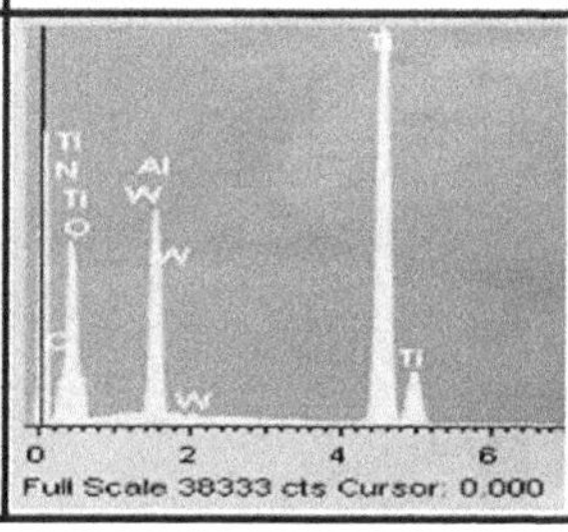

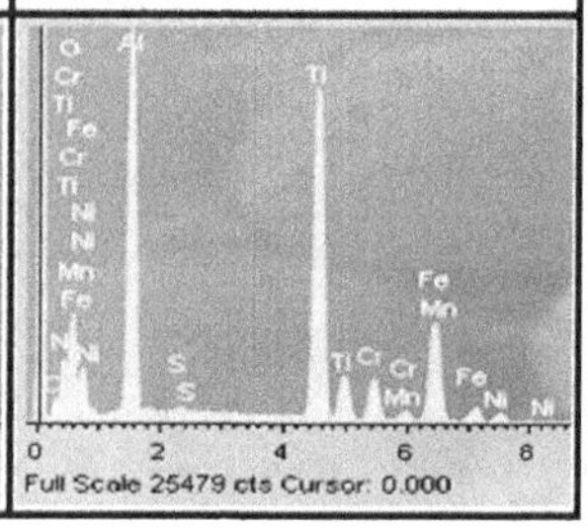

Figure 2.28 EDX chemical composition analysis for new and worn cutting tools

As expected, tool wear is more severe at increasing cutting speed. Besides, wear band in the tool moves from the main edge toward the secondary edge. Figure 2.29 shows that wear is greater in the region near the tool nose radius at a cutting speed of 300 m/min, whereas for cutting speeds of 845 and 870 m/min the most severe wear is observed in the secondary edge. Wear in the secondary edge favors material side flow, as indicated before, and it is predominant at high cutting speeds.

When machining at high speeds, above 300 m/min, lumps of FeCr can be observed welded to the main edge. With increasing cutting speed the lumps tend to displace toward

(a) (b) (c)

(d)

(e)

(f)

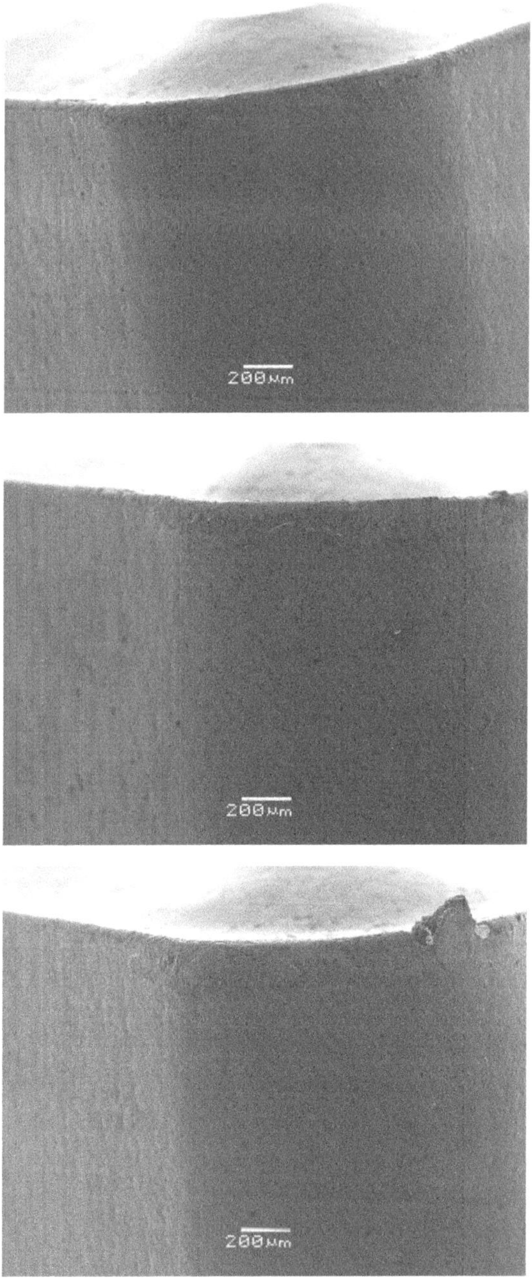

Figure 2.29 Scanning electron micrographs showing tool flank wear at several cutting speeds

the secondary edge, as can be seen in Figure 2.29 for a cutting speed of 870 m/min.

Micrographs of the tool rake face for worn tools and EDX analysis results show a higher level of sulfur (S) and manganese (Mn), indicating the presence of MnS layers, as can be observed in Figure 2.30. A layer of MnS is typical when machining austenitic stainless steels with improved machinability.

At moderate cutting speeds the MnS layer is uniform through the tool edge and it spreads over the tool rake face. This layer acts like a barrier for adhesive wear and it prevents the welding or stiffening of the work material in the tool surface. However, at high cutting speeds the MnS layer formed in the tool rake face is more irregular, as higher temperatures caused by higher cutting speeds makes this layer viscous and unstable. This layer then spreads toward the inner part of the tool, moving away from the edge border. This displacement leaves the cutting edge unprotected and, consequently, welding of the work material takes place on the tool edge border.

In those areas of the tool rake surface not protected by the MnS layer [area B in Figure 2.30(c,d)], a high content of tungsten (W) and low percentages of aluminum (Al), titanium (Ti) and nitrogen (N) are observed. This indicates that tool coating has disappeared in this area and therefore the tool substrate will be in contact with the part material. Under such conditions adhesive and diffusion wear will take place.

From this analysis it is concluded that stability of the protective layer depends on cutting speed. For cutting speeds lower than 450 m/min the layer protects the tool edge; however, stability of the layer cannot be maintained at higher cutting speeds. Its stability is also highly sensitive to geometry and micro-geometry of the tool; therefore, it is

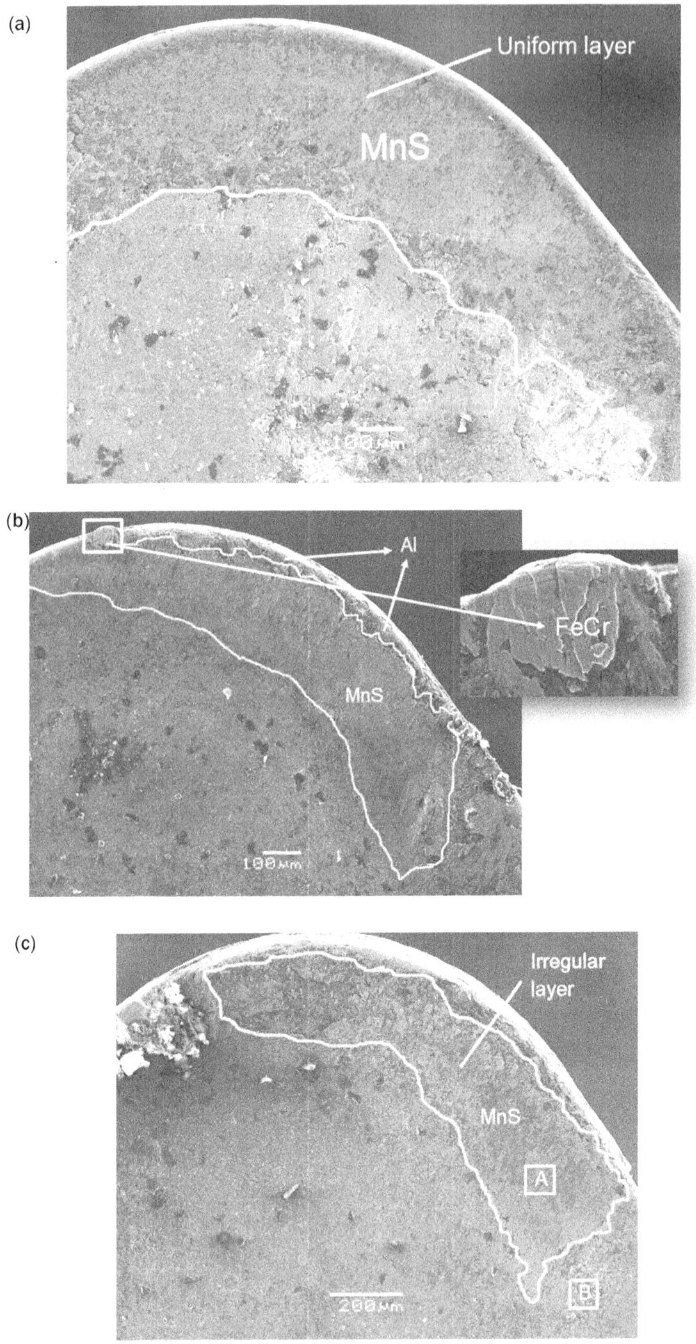
(a)
Uniform layer
MnS
100 µm
(b)
Al
FeCr
MnS
100 µm
(c)
Irregular layer
MnS
A
B
200 µm

(d)

	Zone A		Zone B	
Element	Weight%	Atomic%	Weight%	Atomic%
C K	13.85	26.20	12.74	30.93
O K	32.01	45.44	18.25	33.28
Al K	8.98	7.56	5.57	6.02
Si K	0.29	0.24	-	-
S K	1.20	0.85	1.26	1.15
Ca K	0.19	0.11	0.18	0.13
Ti K	26.36	12.50	21.78	13.26
Cr K	4.24	1.85	4.64	2.60
Mn K	4.94	2.04	2.27	1.20
Fe K	7.12	2.90	14.00	7.31
Co K	-	-	1.51	0.74
Ni K	0.81	0.32	1.58	0.79
W M	-	-	16.21	2.57
Totals	100.00		100.00	

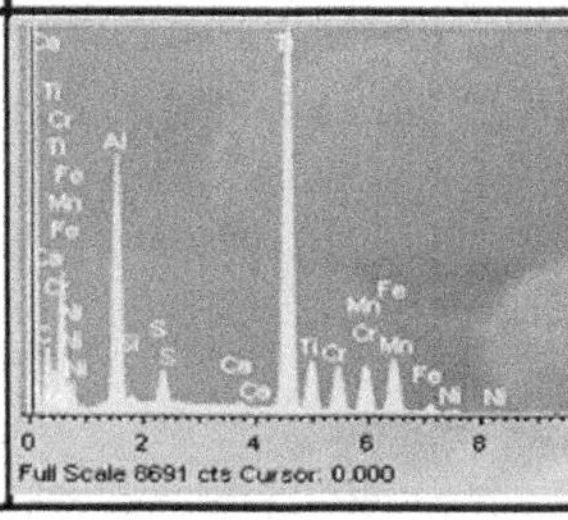

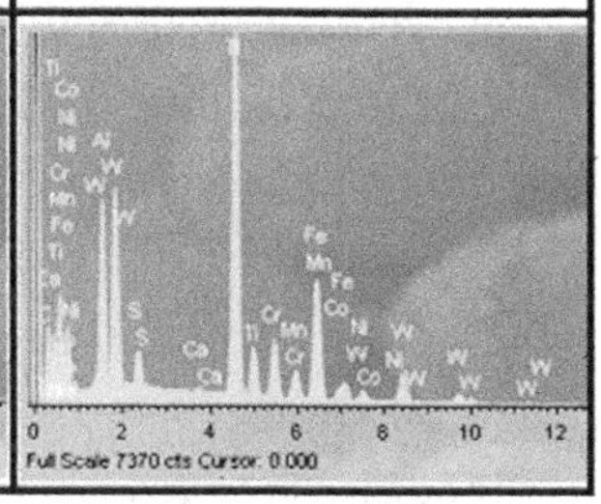

Figure 2.30 Scanning electron micrographs showing crater wear at several cutting speeds: (a) V_c = 37 m/min; (b) V_c = 450 m/min; (c) V_c = 870 m/min. (d) EDX analysis of areas A and B.

necessary to improve tool designs to favor formation of a stable and intense layer along the cutting edge. This will lead to a reduction in tool wear when machining at high cutting speeds.

2.3.3.5 Influence of cutting speed on chip formation

Geometric and morphological analysis of chips was carried out at different cutting speeds. The morphology of a chip, whether continuous or segmented, is determined by material properties and cutting parameters, such as cutting speed, which determine the grade of segmentation (Schulz *et al.*, 2001).

Chips were picked up and analyzed by means of optical microscopy and SEM. In all cases chips were segmented, but geometric differences were observed as a consequence of variations in cutting speed. A segmented chip is formed due to characteristics of austenitic stainless steels, such as high work hardening rate and low thermal conductivity, which prevent dissipation of heat.

Cutting produces strong plastic deformation in the shearing area in a short period of time. Plastic deformation leads to an increase of material strength. At the same time, heat concentrates around the shear zone and produces a thermal softening and decreased yield stress in the material. Moreover, heat generated by deformation does not have sufficient time to be dissipated due to the low thermal conductivity of austenitic stainless steels. In this condition, a local increase of temperature takes place in the shearing plane following a material thermal softening. This effect produces adiabatic bands of deformation that lead to deformation in a nearly planar area, through which sudden slip of segmented chips takes place.

The generated chip is semi-continuous with appreciable tooth size. The chip has wide areas of low deformation and small areas where almost all plastic deformation is concentrated in slip bands. Figure 2.31 shows an image of a segmented chip with two different areas: an area of high plastic deformation, where grains appear severely deformed, and an area of low deformation, where grains are moderately deformed.

SEM images were acquired to analyze geometric differences in chips obtained at different cutting speeds. Ten chip

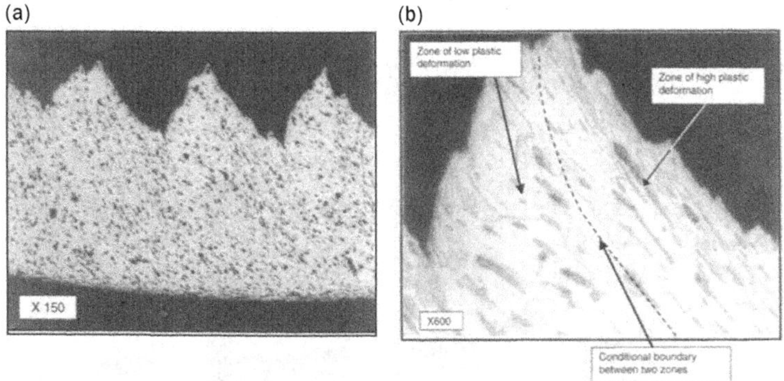

Figure 2.31 **Micrographs of a chip for AISI303 stainless steel: (a) morphology of chip under orthogonal cutting with V_c = 60 m/min and f_n = 0.2 mm/rev; (b) portion of chip**

Source: Astakhov (2006)

samples were considered for each cutting speed so that results were representative. For each chip three images were taken, each one with the chip in an appropriate orientation to measure its thickness and width and to analyze features of the free surface and slip surface (the one in contact with the tool rake face).

Figure 2.32 shows an example of images acquired for each chip at different cutting speeds. The first column shows that the higher the cutting speed the lower the chip length, which is more significant above 600 m/min. This is due to the severe segmentation that occurs at high cutting speeds, chip breaking occurring between lamellae.

The free surface of the chip has a rough aspect. A pattern of lamellae is observed for all cutting speeds as a result of the slip mechanism. Lamellae are smaller and more uniform at moderate cutting speeds, as can be observed in the last column in Figure 2.32. The pattern of lamellae is deeper at higher cutting speeds, indicating high chip segmentation.

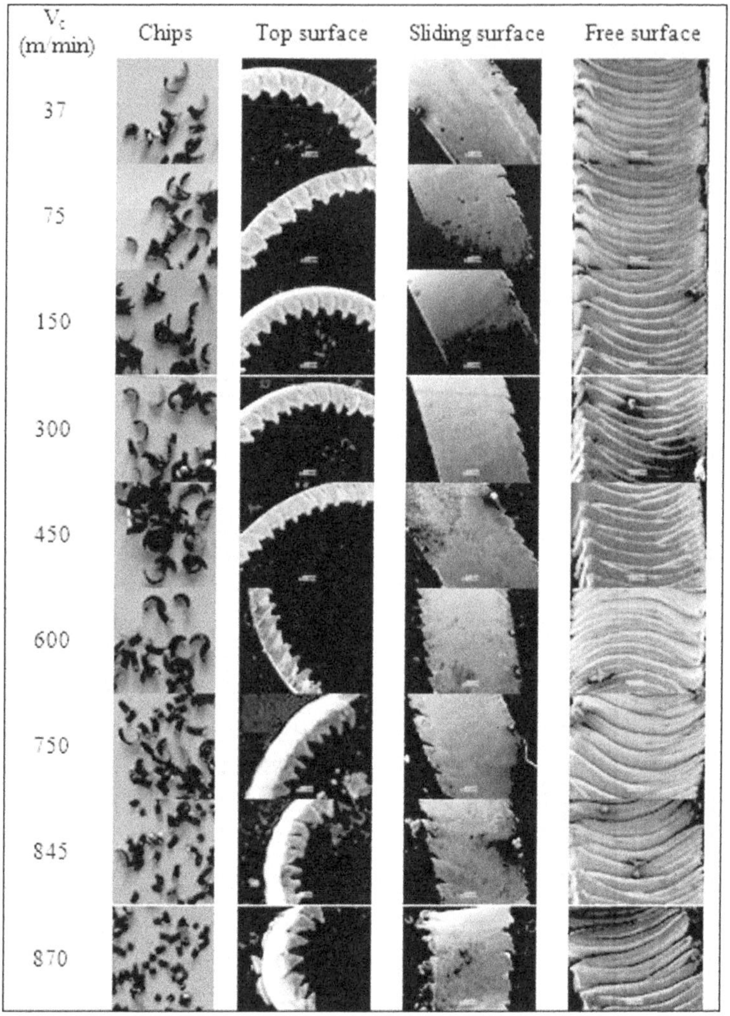

Figure 2.32 Scanning electron micrographs of chips at several cutting speeds

As is well known in the literature, during the cutting process the chip surface that slips on the tool rake is subjected to high pressures and friction forces, reaching very high cutting temperatures of around 1200 °C (Zhang and Guo,

Published by Woodhead Publishing Limited, 2013

2009). The combination of these effects makes this surface flat and bright. Thus, measurements of chip width were carried out in the chip sliding surface. With increasing cutting speed, chip width varies from 1.174 mm (at 37 m/min) to 1.701 mm (at 870 m/min). From these measures it can be concluded that lateral chip flow is occurring, which is higher with higher cutting speed due to high pressures and temperatures in the chip.

It is also observed that chips obtained at moderate cutting speeds (below 300 m/min) have discontinuities in one of the lateral ends. However, at cutting speeds above 300 m/min, cracks and discontinuities appear in both lateral ends (Figure 2.33) due to the severe deformation that occurs at high cutting speeds.

EDX analysis was carried out on sliding surfaces of chips. The results showed the absence of elements of tool coating or tool substrate, indicating that there had been no transference of material from tool to chip, i.e. no diffusion wearing.

Chip thickness (h_c) was measured for each of the cutting speeds. Table 2.8 provides the average values based on ten measurements.

(a)

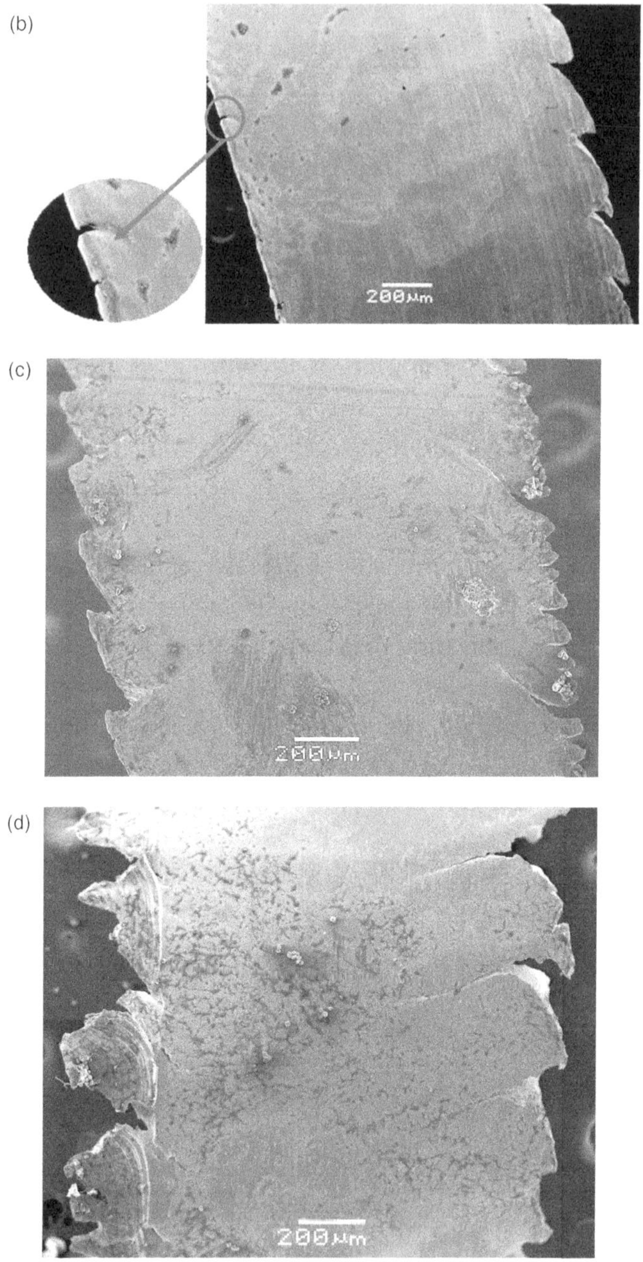

Figure 2.33 **Chip sliding surface at several cutting speeds**

Chip thickness decreases significantly above 450 m/min. In the research carried out by Kishawy and Elbestawi (1999), the authors concluded that if chip thickness reaches a minimum value then material is squeezed between the tool flank face and machined surface, and material side flow is observed. In our study this phenomenon took place when chip thickness was below 0.35 mm at cutting speeds above 450 m/min.

Machinability of a material can be evaluated by means of the chip compression ratio and shearing angle, as these indexes provide important information about the mechanism of material removal and grade of plastic deformation (Thakur *et al.*, 2009). Chip thickness is related to shear angle according to several well-known expressions, such as that stated in the classic Merchant theory:

$$r_c = \frac{h}{h_c}; \qquad \varphi = \tan^{-1}\left(\frac{r_c \cdot \cos\gamma}{1 - r_c \cdot \operatorname{sen}\gamma}\right) \qquad [2.6]$$

Table 2.8 **Chip thickness (h_c), chip compression ratio (r_c) and shearing angle (ϕ) at different cutting speeds**

V_c (m/min)	h (mm)	h_c (mm)	r_c	ϕ (degrees)
37	0.199	0.373	0.5359	29.5593
75	0.199	0.370	0.5403	29.7713
150	0.199	0.451	0.4432	24.8829
300	0.199	0.460	0.4346	24.4310
450	0.199	0.262	0.7630	39.7049
600	0.199	0.253	0.7901	40.7886
750	0.199	0.302	0.6619	35.4273
845	0.199	0.313	0.6387	34.3883
870	0.199	0.254	0.7870	40.6658

Published by Woodhead Publishing Limited, 2013

where r_c is the chip compression ratio, h is non-deformed chip thickness, ϕ is the shearing angle and γ is the tool rake angle. Chip compression ratio (r_c) and shearing angle (ϕ) become higher for lower chip thickness (h_c). High values of ϕ and r_c correspond to lower chip deformation and minor energy, therefore favoring the cutting process.

Table 2.8 provides values for r_c and ϕ calculated from Eq. 2.6. For cutting speeds up to 450 m/min, the chip compression ratio is below 0.5 and approximates to 0.7–0.8 for higher cutting speeds, which indicates that at high cutting speeds the chip suffers less plastic deformation. Shearing angle shows a similar behavior: it remains close to 30° at cutting speeds below 450 m/min and reaches values near 40° when cutting speed is above 450 m/min.

This behavior has already been observed in other materials, such as Inconel 718. Thakur *et al.* (2009) observed that an increase in cutting speed increases the shearing angle, which results in a shorter shear plane. They concluded that if shearing angle increases then thin chips are generated and their speed of formation is higher, resulting in a reduction of cutting forces and friction. Shearing deformations in the primary deformation area diminish with increasing shearing angle.

2.4 Conclusions

In this chapter we have analyzed the behavior of austenitic stainless steels with improved machinability when they are machined under severe cutting conditions (up to 870 m/min) without coolant. The effect of high cutting speed was analyzed through the study of cutting forces, part superficial quality, tool wear and chip geometry. The main reasons for carrying out this analysis at high cutting speed are mainly financial and ecological, in accordance with current trends and technology. It is now considered more advantageous to optimise increasing cutting conditions rather than tool life.

From the results obtained here, a critical cutting speed of 450 m/min is identified. Above this critical cutting speed, behavior of the machined material is different. Traditionally, stainless steels are machined at cutting speeds of 150–300 m/min. However, this research concludes that these materials show good behavior at higher speeds. For example, the main component of cutting forces (F_t) is lower above 450 m/min, which implies less power consumption and less stress and deformation in the cutting tool. Besides, with regard to the machined surface, roughness values at very high cutting speeds are in the same range as for moderate speeds, minimum values for R_a and R_t being at 600 m/min.

There were no cavities or metal debris adhering to the machined surface at high cutting speed; however, material side flow was observed above the critical cutting speed. The depth of the zone affected by micro-structural change increases with cutting speed, although depth values obtained are acceptable (around 600 μm) for all tested cutting speeds. Analysis of chips reveals a significant change in chip geometry above the critical cutting speed of 450 m/min: chip thickness is significantly smaller, which leads to a lower chip compression ratio and higher shearing angle, and therefore less plastic deformation. These parameters are indicative of the good behavior of austenitic stainless steels in dry machining at very high cutting speeds.

Finally, there is an unexplored range of cutting speeds that is of interest for high-performance machining, and at which the behavior of the machined material is very favorable although tool wear is high. Nevertheless, the cost of tool inserts is of secondary importance as compared with other operation costs, for instance hourly machine cost for high-end multitasking machines.

References

Abouelatta O, Mádl J (2001), 'Surface roughness prediction based on cutting parameters and tool vibrations in turning operations', *Journal of Materials Processing Technology*, **118**: 269–77.

Akasawa T, Sakurai H, Nakamura M, Tanaka T, Takano, K (2003), 'Effects of free-cutting additives on the machinability of austenitic stainless-steels', *Journal of Materials Processing Technology*, **143–144**: 66–71.

Astakhov V P (2006), *Tribology of Metal Cutting. Tribology and Interface Engineering. Series, N° 52*. London: B. J. Briscoe. Elsevier.

Chen W (2000), 'Cutting forces and surface finish when machining medium hardness steel using CBN tools', *International Journal of Machine Tools & Manufacture*, **40**: 455–66.

Davim J P (2008), *Machining: Fundamentals and Recent Advances*. London: Springer.

Di Caprio G (1999), *Los aceros inoxidables*. Milán: Ulrico Hoepli Editores.

Dolinsek S, Ekinovic S, Kopac J (2004), 'A contribution to the understanding of chip formation mechanism in high-speed cutting of hardened steel', *Journal of Materials Processing Technology*, **157–158**: 485–90.

Ekinovic S, Dolinsek S, Kopac J, Godec M (2002), 'The transition from the conventional to the High-Speed Cutting region and a chip-formation analysis', *Journal of Mechanical Engineering*, **48**: 133–42.

El-Wardany T, Mohamed E, Elbestawi M (1993), 'Material side flow in finish turning of hardened steel with ceramic tools', *Proceedings of the ASME Winter Annual Meeting*, 159–70, New Orleans.

Fang N, Wu Q (2009), 'A comparative study of the cutting forces in high speed machining of Ti-6Al-4V and Inconel with a round cutting edge tool', *Journal of Materials Processing Technology*, **209**: 4385–9.

Hastings W F, Mathew P, Oxley P L B (1980), 'A machining theory for predicting chip geometry, cutting forces, ets. from work material properties and cutting conditions', *Proceedings of the Royal Society of London. Series A, Mathematical and Physical Sciences*, **371**: 569–87.

He N, Lee T, Lau W, Chan S (2002), 'Assessment of deformation of a shear localized chip in high speed machining', *Journal of Materials Processing Technology*, **129**: 101–4.

ISO 6507-1:1997. *Metallic materials. Vickers hardness test. Part 1: test method.*

Kishawy H, Elbestawi, M (1999), 'Effects of process parameters on material side flow during hard turning', *International Journal of Machine Tools and Manufacture*, **39**: 1017–30.

Liew W Y H, Ngoi B K A, Lu Y G (2003), 'Wear characteristics of PCBN tools in the ultra precision machining of stainless steel at low cutting speeds', *Wear*, **254**: 265–77.

Liew W Y H (2010), 'Low-speed milling of stinless steel with TiAlN single-layer and TiAlN/AlCrN nano-multilayer coated carbide tools under different lubrication conditions', *Wear*, **269**: 617–31.

Lu C, Ma N, Chen Z, Costes J (2010), 'Pre-evaluation on surface profile in turning process based on cutting parameters', *International Journal of Advanced Manufacturing Technology*, **49**: 447–58.

Maranhao C, Davim, J (2010), 'Finite element modelling of machining of AISI 316 steel: Numerical simulation and experimental validation', *Simulation Modelling Practice and Theory*, **18**: 139–56.

Merchant M E (1945), 'Mechanics of metal cutting process', *Journal of Applied Physics*, **16**: 267–324.

Oxley P (1989), *The Mechanics of Machining: an Analytical Approach to Assessing Machinability*. London: Ellis Horwood Ltd.

Pawade R, Joshi S, Brahmankar P, Rahman M (2007), 'An investigation of cutting forces and surface damage in high-speed turning of Inconel 718', *Journal of Materials Processing Technology*, **192–193**: 139–46.

Schulz H, Abele E, Sahm A (2001), 'Materials aspects of chip formation in HSC machining', *Annals CIRP*, **50(1)**: 45–8.

Sullivan K, Wright P, Smith, P (1978), 'Metallurgical appraisal of the instabilities that arise in machining', *Journal of Metals Technology*, **5(6)**: 181–9.

Sun S, Brandy M, Dargusch M (2009), 'Characteristics of cutting forces and chip formation in machining of titanium alloys', *International Journal of Machine Tools and Manufacture*, **49**: 561–8.

Sutter G, Molinari, A (2005), 'Analysis of the cutting force components and friction in high speed machining', *Journal of Manufacturing Science Engineering*, **127**: 245–50.

Thakur D, Ramamoorthy B, Vijayaraghavan L (2009), 'Machinability investigation of Inconel 718 in high-speed turning', *International Journal of Advanced Manufacturing Technology*, **45**: 421–9.

Umbrello D, M'Saoubib R, Outeiro J (2007), 'The influence of Johnson–Cook material constants on finite element simulation of machining of AISI 316L steel', *International Journal of Machine Tools and Manufacture*, **47**: 462–70.

Warnecke G, Bach, P (1988), 'Mechanical and material influences on machined surface in precision turning of steel with ceramics', *Proceedings of the 16th North American Research Conference*, pp. 209–216.

Woei-Shyan L, Chi-Feng L (2001), 'Impact properties and microstructure evolution of 304L stainless steel', *Materials Science Engineering*, **A308**: 124–35.

Yousefi R, Ichida Y (2000), 'A study on ultra-high-speed cutting of aluminium alloy: Formation of welded metal on the secondary cutting edge of the tool and its effects on the quality of finished surface', *Journal of the International Society of Precision Engineering and Nanotechnology*, **24**: 371–6.

Zhang S, Guo, Y (2009), 'An experimental and analytical analysis on chip morphology, phase transformation, oxidation, and their relationships in finish hard milling', *International Journal of Machine Tools and Manufacture*, **49**: 805–13.

3

Forces monitoring in shape grinding of complex parts

A. Di Illio[1]

Abstract: This article deals with the monitoring of process forces when grinding complex mechanical components, such as screw rotors used for the building of gas compressors, which require high dimensional precision and low surface roughness, in order to ensure their proper functionality. Such components must rotate very close together, but without coming into contact, to ensure an airtight seal and to reach maximum efficiency in terms of gas flow rate and compression ratio. This article describes a load force instrumented workpiece holder, employing commercial piezoelectric load washers, to be used for online monitoring of grinding forces. After laboratory calibration, the equipment is then used in a two-level full factorial experimental design for its validation within an industrial environment. The effects of the main grinding parameters (e.g. wheel peripheral speed, feed rate and depth of cut) on the evolution of grinding forces during the process are investigated. The methodology used here can be extended to the grinding of helical surfaces of any shape, such as in precision manufacturing of helical gears.

Key words: Force monitoring, screw rotor, shape grinding process.

[1] University of L'Aquila, Italy.

Published by Woodhead Publishing Limited, 2013

3.1 Introduction

Compressors are operating machines used to enhance the pressure of a fluid by supplying mechanical energy to it, without any notable external heat exchange due to the brief crossing time of the fluid inside the machine. Screw compressors belong to the category of volumetric compressors, where gas compression is achieved through a reduction of volumes enclosed between the conjugated surfaces of the teeth of two rotors. Their increased use in industry has been due to a number of advantages such as high reliability and long technical life, high velocity and flow rate, high compression ratios (up to 16), simple maintenance and low operating costs (Stosic *et al.*, 2005). Conversely, there are high manufacturing costs associated with the rotors due to the tight tolerances required.

The manufacturing of screw rotors includes two main phases, i.e. milling to produce the helical grooves and grinding to give the profile the required tolerances, which are of the order of microns. Several research groups have investigated production issues in recent years, dealing with the creation of helical groves by milling (Xiao *et al.*, 1996; Katsumi *et al.*, 2001; Yao *et al.*, 2005) and the subsequent finish grinding. Stosic *et al.* (2001, 2005) and Stosic (2006) studied functional aspects and the related problems of manufacturing, particularly due to tool wear, and have suggested several strategies to make tool wear uniform.

To bring the rotor geometry to its final tolerances, the grinding operation requires particular attention. Grinding of screw rotors belongs to the category of shape (or form) grinding, which, among such abrasive removal processes, is particularly complex.

In fact, to achieve the required precision along the entire extension of the helical grooves, only very low wheel wear is tolerated, requiring frequent profiling and dressing operations of the grinding wheel. Frequent stops of the machining for wheel regrinding would introduce further uncertainty to the overall precision of the workpiece and, in addition, reduce the productivity of the process.

On the other hand, low grinding wheel wear is achieved by using profilable ceramic wheels made of superabrasive materials, such as CBN (cubic boron nitride), together with an optimized grinding cycle from roughing to finishing passes. The economy of the machining is highly influenced by such choices, as superabrasive grinding wheels are very expensive and throughput of the process can be very slow, due to the necessity of using parameters which keep wear low; therefore, the overall operation cycle must be optimized to enhance productivity and reduce costs. Such optimization can be obtained by in-depth study of the process to allow assessment of the proper machining parameters. On the other hand, optimization of the grinding cycle may not be sufficient if tight control of the process trend is not performed, so as to overcome unpredictable situations encountered by the tool, which could depend on inhomogeneities in the material or dimensional imprecisions left by the previous grooving operation performed by the milling machine.

Among grinding operations, shape grinding is particularly complex not only due to geometrical complications, but also from the point of view of control, given critical aspects regarding wear and thermal effects, which can greatly affect the final quality of the workpiece.

Regarding shape grinding, Malkin (1989) investigated the grinding operation by rotating the wheel axis with respect to the workpiece axis, extending the analysis to arbitrary wheel profiles. Based on work by Graham and Falconer (1979), Malkin showed that despite the increased complexity, some relationships derived for two-dimensional plunge grinding could still be applied by making the necessary geometrical corrections, obtaining an expression for the cutting path of the grains and maximum chip thickness.

Numerous authors have investigated the shape grinding of gears. Weinert *et al.* (2007) report an optimization method based on simulation for NC-shape grinding, while You *et al.* (2003) propose a mathematical model for profiling grinding wheels for gears. An experimental analysis aimed to develop models for NC-shape grinding has been provided by Biermann *et al.* (2008). Thermal alterations to the material in zones where heat cannot be dissipated easily represent a real concern in such cases.

Most researchers stress the application of CBN in shape grinding, as this material has the best properties to work well with these applications. In fact, CBN can be used for a long period of time before requiring dressing, and cracks and thermal alterations are minimal thanks to its high thermal conductivity (Kopac and Kranjnic, 2006; Pavel and Srivatsa, 2007). Inoue *et al.* (1998) investigated the effects of CBN grinding on carburized gears and found an increase in fatigue strength thanks to increasing compressive residual stresses.

Seeded gel grinding wheels, containing microcrystalline aluminum oxide, also have good potential in shape grinding. Eranki *et al.* (1992) carried out a comparison of the performance of grinding wheels containing newly developed 'seeded gel' aluminum oxide abrasive with wheels containing conventional monocrystalline aluminum oxide, and reported a higher removal rate and grinding ratios when using the former.

A constant concern when dealing with grinding operations relates to vibrations, due to grinding forces and the elastic characteristics of the system comprising wheel, workpiece and grinding machine, which cause poor tolerances and surface finish (Alfares and Elsharkawy, 2000).

Guo *et al.* (2003) proposed an optimization method for continuous dressing aimed at reducing cycle time and wheel wear using an adaptive workpiece speed and infeed, based on grinding models and power monitoring.

Regarding compressor rotors, Wei and Zhang (2010) report an analysis concerning the profiling of CBN wheels to give high precision in grinding screw rotors.

Here we focus on the finish operation by grinding. Our motivations lie in the particular complexity of the surfaces being ground (shape grinding) with very high precision, over a surface which extends along the entire length of the component. Our objective is to gain insight into the process, and hence to optimize the cycle to enhance the quality and reduce production times as well as to explore the possibility of continuously monitoring the grinding forces during the process. To this aim, an instrumented workpiece holder has been designed and realized to monitor grinding forces during the process. The device has been previously calibrated in the laboratory, and has been used in a real manufacturing environment to assess its capability in capturing grinding features that reveal any malfunctioning in the process, in order to alert the operator and allow them to act promptly.

3.2 Generality of screw rotors

Essentially, a screw compressor consists of two rotors, one male and one female, rotating in opposite directions, which enclose, compress and transfer a gas volume from inlet to outlet. There are different configurations of crew compressors, both symmetric and asymmetric. Figure 3.1 shows a pair of asymmetric rotors.

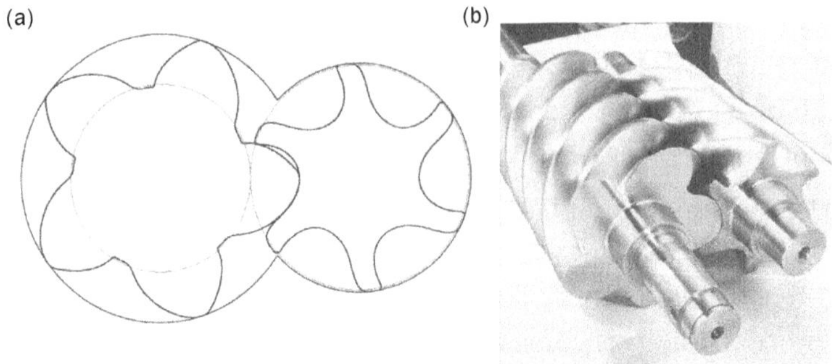

Figure 3.1 A pair of asymmetric rotors. (a) Schematic drawing, (b) finished parts. Left, male; right, female

The functional parameters used to classify such compressors are gas flow rate, velocity and compression ratio. The most important requirement is that the geometry is such that the cross-section for fluid processing is as high as possible and loss by leakage is minimal. According to the compressor operating field, rotors can be realized with a different number of lobes and, with reference to the cross-section, they may exhibit symmetric or asymmetric profiles. During functioning of the compressor, these profiles must be as close as possible to avoid gas loss, but without entering into contact with each other, in order to avoid wear or gripping. In some cases, sealing oil can be used to enhance the airtightness if, for a particular application, contamination of the gas due to the contact with the oil does not represent a concern.

The grinding of such rotors requires the adoption of a sophisticated machining cycle which starts with the measurement of the helix deviation by means of a gear measuring machine, to reveal the thickness of the stock to be removed, and the subsequent precise positioning of the workpiece between headstock and tailstock in the grinding machine. There are different grinding strategies to plan the machining. One example includes up to four sub-cycles; the first is employed to bring the workpiece to a reference geometry. This is followed by removal of the stock in one or more roughing runs, one sub-cycle for semi-finishing and one cycle for final finishing. The entire machining cycle can include a number of dressing cycles aimed to restore the correct geometry of the wheel and to renew its grinding ability by exposing fresh grains on its active surface. Definition of the correct machining parameters and the above-mentioned sequence, included dressing, can greatly influence the quality of the workpiece, tool life, machining time and hence costs.

3.3 Workpiece holder

By taking into account that mounting of the workpiece on the grinding machine is performed between headstock and tailstock, the workpiece holder has been thought to fulfil industrial requirements.

3.3.1 Design

A tailor-made workpiece holder was designed to mount the workpiece through two piezoelectric load washers by Kistler Instruments (Hook, UK), each one having two load channels, namely model 9063 for F_x and F_y, and 9065 for F_z and M_z. Piezoelectric sensors were chosen for their high rigidity, which does not affect the elastic characteristics of the mounting system. The design of the workpiece holder was made based on a balance between the need to keep the overall dimensions low and the need to not reduce the rigidity of the workpiece-holding system. Figure 3.2(a) shows two load washers based

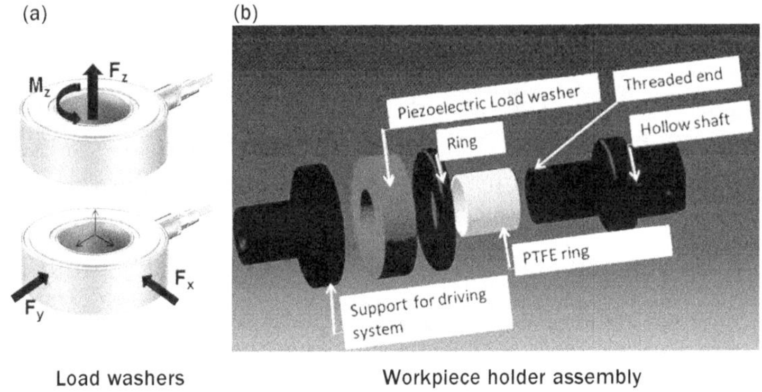

Figure 3.2 Schematic of the workpiece holder assembly

on piezoelectric sensing elements; the upper one is sensible to torque and axial force, while the lower one is sensible to two in-plane force components, i.e. F_x and F_y. An exploded assembly scheme is shown in Figure 3.2(b). The hollow shaft on the right side allows for the connection to the shaft of the rotor by means of six centering screws. On the left side of the hollow shaft a PTFE (polytetrafluoroethylene) muff is inserted, allowing the centered mounting of a steel ring and the two load sensors which are held tightly together by screwing the threaded end of the hollow shaft into the support to drive the system shown on the left. In such a manner, the two load rings, whose outputs were continuously monitored during the assembling operation, were preloaded up to 1.8 kN (which was much lower than F_z full scale of 120 kN). Such preload allowed us to measure both tensile and compressive axial forces under a wide range of conditions for the present application and ensured transmission of the tangential force necessary for measurement of the torque M_z and radial forces F_x and F_y.

Note that the instrumented workpiece holder represents an elastic system whose single parts sustain force components according to their rigidity. Therefore, the sensitivity of the load washers mounted in such an assembly was inevitably reduced due to the fact that any load component applied to the workpiece mounted on it is invariably absorbed by the mechanical parts comprising the assembly. In particular, the cross-section and length of the hollow shaft which fit together and clamp the two ring sensors play an important role.

3.3.2 Calibration

Calibration along the *z* direction (workpiece axis) was made simply by loading the workpiece holder, positioned vertically, with different weights along its axis.

Calibration of torque, F_x and F_y components was made by mounting a real workpiece on the instrumented holding device and positioning the assembly between a headstock and a tailstock of a calibration bench, as in a machining operation (Figure 3.3).

The load scheme for such a mounting system is a beam supported at its ends that is subjected to a load for unit length, due to its own weight, varying according to the local cross-section, which during the machining is subjected to a concentrated force exerted by the grinding wheel. This force exhibits components along axes *x*, *y* and *z* and generates a torque. In such a scheme, F_x and F_y revealed by the sensor represent the components of the shear force acting on the

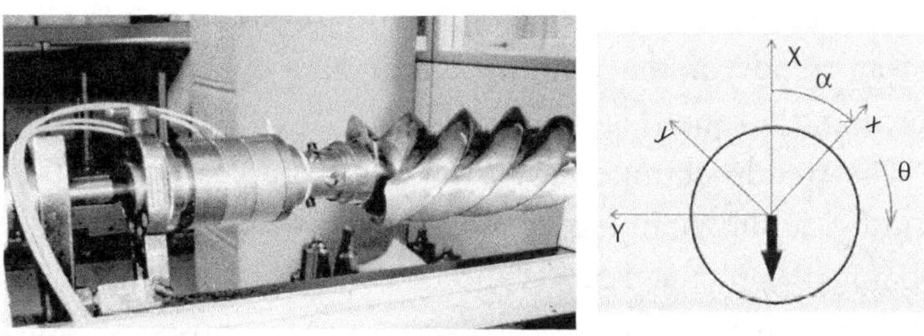

Figure 3.3 Calibration scheme for radial forces and torque

beam, which, during machining, varies due to the axial displacement of the application point of the resultant grinding force. Therefore, the values obtained during calibration by positioning different weights at the same distance from the sensor should be corrected to take this into account.

A further consideration relates to what happens when the workpiece rotates, as it does during machining. In fact, the components detected by the sensor vary according to the rotation angle and, in addition, depend on the initial orientation (α in Figure 3.3) of the sensor with respect to a reference direction (x in Figure 3.3). Assuming that, prior to calibration, the initial signal given by the sensor is reset to zero, the output signals for components F_x and F_y vary as the workpiece rotates according to the following relationships:

$$F_x = -P\cos(\theta + \alpha) + P\cos\alpha \qquad [3.1]$$
$$F_y = -P\sin(\theta + \alpha) + P\sin\alpha$$

where the angle α takes into account that the sensor cannot be mounted exactly with the same orientation and θ is the rotation angle during machining.

For the torque signal, the calibration was made by applying different weights at the radius of the shaft through a fine wire.

The results obtained from the calibration procedure show linear trends of the output voltage as a function of axial force and torque (Figure 3.4).

The trends obtained for the radial force components F_x and F_y are shown in Figure 3.5(a), while the theoretical ones, obtained applying Eq. 3.1, are shown in Figure 3.5(b). As can be seen the trends reflect quite nearly those calculated theoretically.

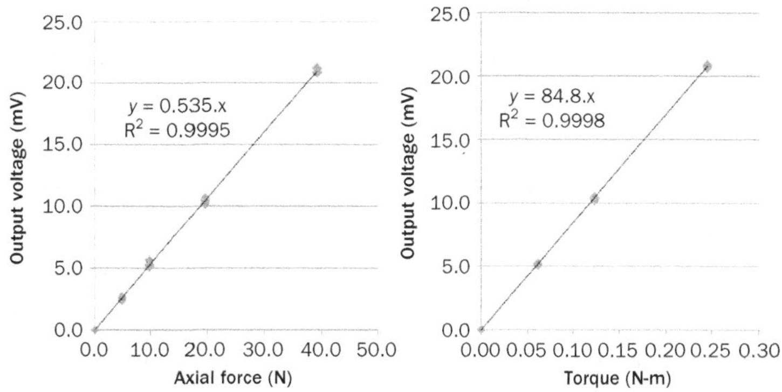

Figure 3.4 Calibration curves for axial force and torque

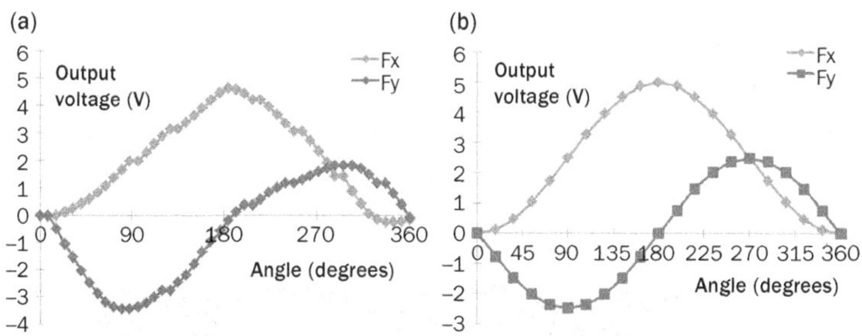

Figure 3.5 Force trends as a function of the workpiece rotation angle, when the *x*-axis of the sensor is oriented vertically: (a) experimental, (b) theoretical

3.4 Experimental tests

Experimental tests were carried out to assess the suitability of the workpiece mounting set-up in an industrial environment and to perform investigations on the grinding forces and their evolution during the grinding cycle.

The experimental set-up is shown in Figure 3.6. The workpiece is mounted in a grinding machine between headstock and tailstock; note that the signal cables (arrowed) coming from the sensors do not interfere with the grinding wheel. The axial force employed to clamp the workpiece was set at 245 N.

Such a set-up is possible because the workpiece length corresponds to no more than one pitch of the helical grooves. Therefore, during the grinding operation, it performs slightly more than one full rotation, while the grinding wheel head moves by one pitch. Thereafter the rotor inverts its rotation sense while the grinding head returns to its initial position.

The measuring chain is completed by a model 5017B charge amplifier (Kistler Instruments) and an analog-to-digital card (model PMCIA NI DAQCard-6024; National Instruments). Prior to tests starting, all signals produced by the sensors due to the mounting forces were zeroed.

Figure 3.6 Workpiece mounted in a grinding machine

To investigate the possible effects of clamping forces, weight of the assembly formed by the workpiece and the instrumented equipment as well as the grinding coolant jet, a series of idle strokes, without material removal, were first conducted. In particular, idle tests with and without coolant were performed and compared.

The grinding process parameters chosen for validation of the equipment were wheel peripheral speed, feedrate of the workpiece along its axis, and depth of cut measured from the point of contact between the wheel and the rotor at the beginning of the operation. A two-level full factorial design of experiment with three parameters was adopted, for a total of eight different experiments, each one repeated three times. Low and high parameter values, as detailed in Table 3.1, were chosen on the basis of standard industrial practice with this machining operation. The grinding wheel adopted for tests, having a maximum external diameter of 400 mm, was a seeded gel wheel, type 3SG60H12VXPC (Norton), containing 30 per cent microcrystalline ceramic abrasive.

Each test consisted of two runs during which the rotor performed one counterclockwise rotation with the peripheral speed in the same direction as the workpiece speed (down grinding) and one returning to its initial position by rotating in the reverse direction (up grinding). After each test the grinding wheel was subjected to a profiling cycle, following the manufacturer's procedures, in order to maintain its cutting abilities.

Table 3.1 Experimental factor levels employed for the full factorial experimental design

Parameter	Low level	High level
Wheel speed (m/s)	32	38
Axial feedrate (mm/min)	1500	3000
Depth of cut (mm)	0.05	0.10

3.5 Results and discussion

Trends of the three force components and torque, F_x, F_y, F_z and M_z, respectively, as a function of workpiece rotation are shown in Figure 3.7. Each diagram contains a graph for an idle stroke when the coolant nozzle was closed, and one for the case when the coolant nozzle was opened. Both F_x and F_y follow trends similar to those reported in Figure 3.5, but shifted along the abscissa. Each curve extends for about 380°, i.e. more than one revolution, which is the workpiece rotation required in each half cycle to grind the workpiece along its length. It is simple to demonstrate that these components can be still represented by Eq. 3.1 at a value of α of about 105°, which was approximately the angle exhibited by the load sensor, with respect to the reference orientation assumed in Figure 3.5, when the workpiece started to rotate. In fact, it was almost impossible to mount the workpiece on the holding support with the x-axis of the

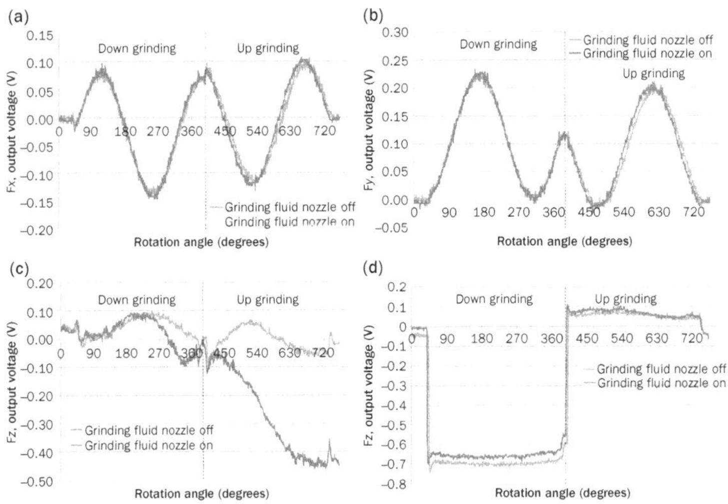

Figure 3.7 **Force components and torque during idle strokes without material removal**

load sensor set as during the calibration test in the laboratory. However, this does not represent a real drawback.

It can be seen that components F_x and F_y are almost independent of the presence of the coolant jet. This is not so for the axial force F_z, which exhibited marked differences in the two cases. This behavior was attributed to the force generated by the static pressure of the fluid hitting the helical grooves of the workpiece and the surface of the tailstock. The jet generates a different axial force when the coolant flow hits the workpiece with a tangential component having the same direction as the workpiece velocity (case of down grinding) with respect to the opposite case (up grinding), which occurs in the return stroke. In addition, some oscillations, which occurred with repeatability in each test, are also present in the absence of coolant. This could be attributed to mechanical imperfections in the driving system of the machine tool, which caused small periodic variations of the axial force between the tailstock and the headstock when the workpiece rotates.

Regarding torque, Figure 3.7(d) shows the torque values required to rotate the workpiece in idle conditions, which exhibits very different values when the workpiece is rotated in one sense or in the other. When the workpiece is rotated counterclockwise (down grinding) the torque is markedly higher than that measured in the opposite condition, although this behaviour may differ from one grinding machine to another. Figure 3.7(d) shows that driving torque is largely independent of coolant effects.

Figures 3.8 and 3.9 show the force components and torque for two different grinding conditions, one of the lightest and the heaviest, respectively, among those chosen for the experimental design. The combination of high values of feedrate and depth of cut, together with a low value of wheel peripheral speed, produces the highest value of

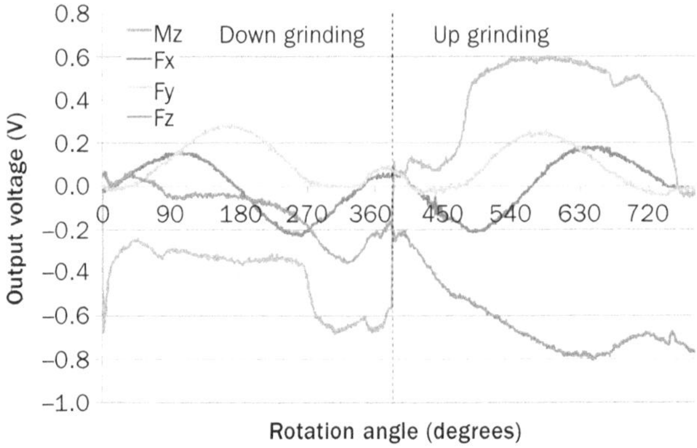

Figure 3.8 Trends of force components and torque as a function of the rotation angle of the workpiece: wheel speed = 32 m/s, feedrate = 1500 m/min, depth of cut = 0.05 mm

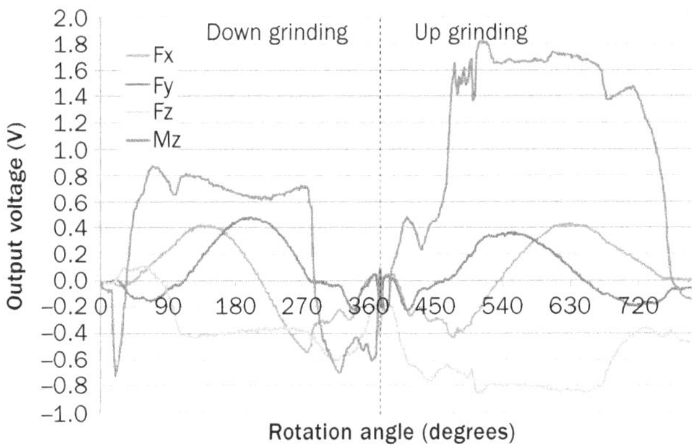

Figure 3.9 Trends of force components and torque as a function of the rotation angle of the workpiece: wheel speed = 32 m/s, feedrate = 3000 m/min, depth of cut = 0.1 mm

removed chip thickness, and hence the highest machining forces. In the first case (Figure 3.8), it can be seen that torque is still negative in the down grinding part of the cycle, a value that is smaller than that required to overcome the friction resistance to put the workpiece into rotation. Taking into account the values measured in the idle tests, we estimated a maximum torque of about 0.7 Nm in the down grinding part of the cycle and about 1.0 Nm in the return stroke. The shear components of the grinding forces, namely F_x and F_y, continue to exhibit trends similar to those observed in idle strokes.

When the axial feedrate of the workpiece and depth of cut increase (Figure 3.9), the above parameter trends maintain their characteristics, but with increased values. In particular, maximum torque values of about 3.0 Nm, in both outward and return strokes, can be estimated.

Regarding F_z, it is seen to be dependent on the pressure exerted by the coolant jet, whose maximum effect remains almost unchanged for both cases reported in Figures 3.8 and 3.9.

At higher cutting speed (Figure 3.10), there is a general decrease of all maximum force values as well as torque; this is in line with the reduction of chip thickness to which the cutting velocity gives rise.

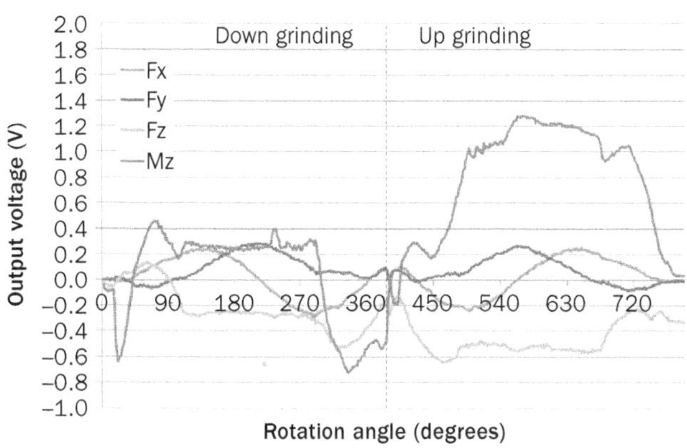

Figure 3.10 **Trends of force components and torque as a function of the rotation angle of the workpiece: wheel speed = 38 m/s, feedrate = 3000 m/min, depth of cut = 0.1 mm**

Our experimental design has allowed us to show the suitability of the instrumented workpiece holder to discriminate the effects of grinding parameters and therefore to assess its utility in building a surface response aimed at optimizing the process.

For brevity, only the maximum torque values have been considered to examine the main effects of the grinding parameters. Such main effects (Figure 3.11) show trends that are in agreement with what is generally expected in a machining process, as already discussed. In particular, it can be seen that an increase of wheel peripheral speed decreases the torque, while both feedrate and depth of cut produce an increase. These trends are in agreement with chip thicknesses removed by the tool under different situations.

The variations highlighted have shown that the system is sufficiently sensitive to variations in the range of grinding parameters employed in the present study.

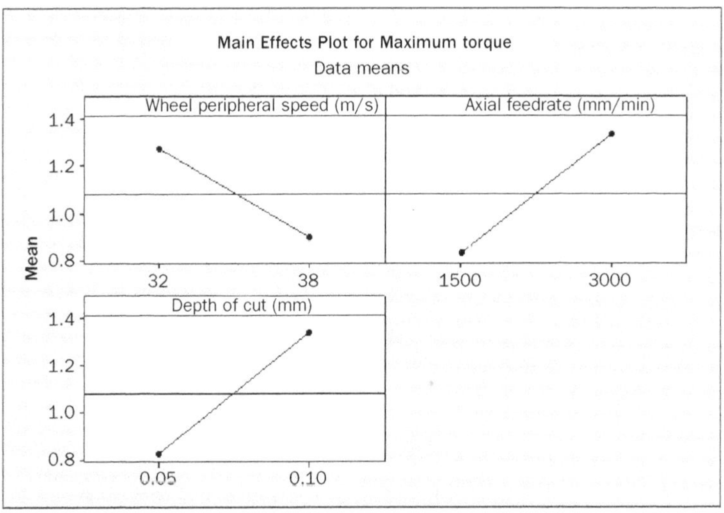

Figure 3.11 Main effects plot of the grinding parameters for torque

3.6 Conclusions

A workpiece holder system, instrumented to monitor the forces during shape grinding of complex surfaces, such as screw rotors for compressors, has been realized and tested in a real production environment. The system has shown good responses in terms of sensitivity to the main process variables and proved to be suited to work in a real production machine. The experimental tests carried out under different process conditions revealed trends of the main grinding forces over the machining cycle. The trends of maximum forces registered in one cycle, found by implementing a two-level experimental design, agree with what is generally found in grinding operations.

The system as realized, conveniently engineered, can find application for cycle optimization and control aimed to reduce the production time in shape grinding of complex surfaces, which require tight tolerances and high surface finish.

Acknowledgements

The author would like to thank Eng. Vittorio Borghetti for having carried out the tests and Samputensili S.p.A. for providing the grinding machine tool and for assistance in the experimental tests.

References

Alfares M, Elsharkawy A (2000), 'Effect of grinding forces on the vibration of grinding machine spindle system', *International Journal of Machine Tools and Manufacture*, 40: 2003–30.

Biermann D, Blum H, Jansen T, Rademacher A, Scheidler A, Weinert K (2008), *Experimental Analyses to Develop Models for NC-Shape Grinding with a Toroid Grinding Wheel.* Hannover: Denkena, B., pp. 279–87.

Eranki J, Xiao G, Malkin S (1992) 'Evaluating the performance of seeded gel grinding wheels', *Journal of Materials Processing Technology,* **32**: 609–25.

Graham W, Falconer D (1979), 'Wheel-workpiece conformity in form grinding', *Proceedings of the Nineteenth International Machine Tool Design and Research Conference*, p. 615.

Guo C, Campomanes M, Mcintosh D, Becze C, Green T, Malkin S (2003), 'Optimization of continuous dress creep-feed form grinding process', *CIRP Annals - Manufacturing Technology*, **52**(1): 259–62.

Inoue K, Sonoda H, Deng G, Yamanaka M, Kato M (1998), 'Effect of CBN grinding on the bending strength of carburized gears', *Journal of Mechanical Design,* **120**: 606–11.

Katsumi K, Kazumasa Z, Hisashi T (2001), 'Gear-cutting tool for screw-compressor rotors (built-up hob)', *JSME International Journal. Series C, Mechanical Systems, Machine Elements and Manufacturing*, **44**(3): 802–7.

Kopac J, Kranjnic P (2006), 'High performance grinding – a review', *Journal of Materials Processing Technology,* **175**: 278–84.

Malkin S (1989), *Grinding Technology – theory and application of machining with abrasives.* Ellis Horwood Limited.

Pavel R, Srivatsa A (2007), 'An experimental investigation of temperatures during conventional and CBN grinding', *The International Journal of Advanced Manufacturing Technology*, **33**: 412–18.

Stosic N (2005), Tool wear and its compensation in screw rotor manufacturing, available at: www.city.ac.uk/sj 631.

Stosic N (2006), 'A geometric approach to calculating tool wear in screw rotor machining', *International Journal of Machine Tools and Manufacture*, **46**: 1961–5.

Stosic N, Smith I K, Kovacevic A (2001), 'Calculation of rotor interference in screw compressors', *Proceedings of the 3rd International Compressor Technique Conference*, Wuxi City China, August, pp.134–42.

Stosic N, Smith I K, Kovacevic A (2005), *Screw Compressors, Mathematical Modelling and Performance calculation*. Berlin: Springer.

Wei J, Zhang G (2010), 'A precision grinding method for screw rotors using CBN grinding wheel', *The International Journal of Advanced Manufacturing Technology*, 48: 495–503.

Weinert K, Blum H, Janse T, Rademacher A (2007), 'Simulation based optimization for the NC-shape grinding process with toroid grinding wheel', *Production Engineering Research Development*, 1: 245–52.

Xiao D Z, Li K B, Wang Z Q, Liu D M (1996), 'Computer aided design software package for conjugate helical surfaces', *Journal of Materials Processing Technology*, 61: 72–7.

Yao L G, Ye Z H, Dai J S, Cai H Y (2005), 'Geometric analysis and tooth profiling of a three-lobe helical rotor of the roots blower,' *Journal of Materials Processing Technology,* 170: 259–67.

You H Y, Ye P Q, Wang J S, Deng X Y (2003), 'Design and application of CBN shape grinding wheel for gears', *International Journal of Machine Tools and Manufacture*, 43: 1269–77.

4

Optimization of minimum quantity lubrication in grinding with CBN wheels

E. C. Bianchi,[1] *P. R. Aguiar,*[1]
R. C. Canarim[1] *and A. E. Diniz*[2]

Abstract: The application of minimum quantity lubrication (MQL) to metal machining has emerged as an alternative means to reduce cutting fluid usage, leading to cleaner production. In grinding, although MQL is considered to be progress regarding the environment, its application is restricted due to excessive heat generation and wheel clogging, which reduce final product quality and increase tool wear. This chapter reviews the research using MQL grinding aimed at improvements in terms of both wheel cleaning (by using an additional compressed air jet) and cutting fluid reduction (by adding water to the MQL oil).

Key words: CBN, cubic boron nitride, grinding, minimum quantity lubrication, MQL, wheel cleaning.

4.1 Introduction

Cutting fluids play an important role in machining, as they usually extend tool life and provide high workpiece quality (Diniz *et al.*, 2008).

---

[1] UNESP, Brazil – Univ. Estadual Paulista.

[2] UNICAMP – State University of Campinas, Brazil.

Published by Woodhead Publishing Limited, 2013

However, some drawbacks emerge from their application, such as: high costs related to purchase, handling, maintenance and disposal, and environmental and health hazards (Alves *et al.*, 2011). According to Tawakoli *et al.* (2007), this search to reduce cutting fluids is now of great importance. However, the output variables that characterize the final quality of a ground workpiece must not be impaired by the new techniques. They need to be equally or more efficient in terms of workpiece quality and cost reduction, minimizing environmental impact, but also maintaining the high-quality standards associated with grinding.

Among the techniques which promote the reduction of cutting fluids are dry machining and minimum quantity lubrication (MQL). Nevertheless, their results are quite different depending on the machining process, due to the different tool/chip contact lengths, different tools and other factors inherent to each process. In this aspect, abrasive machining processes require more cooling-lubrication, which makes dry machining non-viable in some conditions (Weinert *et al.*, 2004).

Grinding is an abrasive machining process in which material is removed by interaction of the abrasive grain with the workpiece (Ren *et al.*, 2009). According to Diniz *et al.* (2008), grinding is a standard finishing process used in the manufacturing of high-quality components, in terms of shape and surface finishing (geometric and dimensional tolerances). Also, conventional cutting fluid application systems are relatively inefficient in the grinding process, particularly under severe conditions (Catai *et al.*, 2008).

Silva *et al.* (2007) and Tawakoli *et al.* (2010) studied the application of MQL in external cylindrical grinding of SAE 4340 and DIN 100Cr6 steels, with a conventional alumina wheel, and their results for MQL showed similar or superior performance, when compared with conventional cooling-lubrication (flood coolant).

Sadeghi *et al.* (2009), on the other hand, found that a reduction in cutting fluid usage impairs wheel cleaning, leading to clogging of its pores, which decreases the cutting capacity of the abrasive tool.

Similarly, Alves (2007) studied the influence of cooling-lubrication methods when grinding SAE 4340 steel using cubic boron nitride (CBN) wheels, and showed that MQL provided lower performance with regard to surface roughness, roundness errors and diametric wheel wear, compared with the conventional method (flood coolant).

The present chapter reviews some of the results concerning improvements of MQL in CBN grinding, in terms of wheel cleaning and application of MQL with addition of water.

4.1.1 Motivation

The problems associated with the application of cutting fluids have led both industry and research centers to seek new methods to reduce their usage and drawbacks (Fratila and Caizar, 2011). In this regard, MQL is as an excellent alternative, providing economic and environmental improvements (Min *et al.*, 2005).

However, this technique requires further improvements in order to achieve the same finishing quality obtained with flood coolant application. Wheel clogging, as described by Sadeghi *et al.* (2009), is one of the main concerns.

Cameron *et al.* (2010) studied the application of a secondary cutting fluid jet, in order to promote wheel cleaning when using flood coolant. They noted that the reduced wheel clogging minimized grinding specific energy.

4.2 Literature review

4.2.1 Problems with cutting fluid usage

Cutting fluids are used to reduce friction, through lubrication, which is a tribological phenomenon characteristic of machining processes, and occurs at the contact between workpiece and tool. Also, cutting works to remove the heat generated from the cutting zone through cooling (Pawlak *et al.*, 2004).

Stanford *et al.* (2007) and Tawakoli *et al.* (2007) stated that cutting fluids also hinder corrosion of both workpiece and machine tool, remove the machined chips from the cutting zone and provide wheel cleaning.

Due to the intrinsic nature of grinding, being an intense heat generation process, large amounts of cutting fluid are used, aiming to reduce the high temperatures at the contact zone (Alves *et al.*, 2009). Cutting fluid composition as well as nozzle design and placement are parameters which greatly influence productivity, workpiece final quality and wheel wear (Brinksmeier *et al.*, 1999).

According to Irani *et al.* (2005) cutting fluids carry away small particles after application, such as wheel grains, machined chips and other impurities. Therefore, due to contamination, all the cutting fluid must at some time be replaced and disposed of. However, this disposal is frequently conducted inappropriately, thus harming the environment.

One of the main problems of machining processes, especially grinding, is related to the use of cutting fluids. This is due to the considerable damage that these substances cause both to human health and to the environment (Bartz, 1998). The main cause of these concerns about cutting fluids lies in their chemical composition. Components such as anti-corrosives, anti-foaming agents and biocides are sometimes added, the last-named being one of the most hazardous additives to machine operators (Hong and Broomer, 2000).

The most important problems concerning cutting fluids were discussed by Howell *et al.* (2006). Damage to the skin, digestive and respiratory systems can be caused due to contamination by heavy metals. Contact with the skin can cause allergies and dermatitis, depending on fluid type and exposure time. Digestive and respiratory illnesses can occur due to ingestion of small particles and the breathing in of cutting fluid mist.

Besides these environmental and health problems, there are also economic concerns related to the use of cutting fluids, as this represents a significant portion of the manufacturing costs (about 15–17 per cent) (Byrne and Scholta, 1993). This is because, besides purchase costs, fluid maintenance is necessary to preserve its properties. Thus, it is important, among other things, to promote temperature monitoring, water treatment, fluid circulation throughout the machine tool and pH control.

According to Ebbrell *et al.* (2000), despite the great advantages brought about by the use of cutting fluids, frequently its application is inefficient, with considerable losses. It is important to ensure the effective penetration of cutting fluid in the interface between workpiece and tool, improving the final quality and reducing costs.

Dhar *et al.* (2006) stated that, despite the technological advantages provided by cutting fluids, their drawbacks are being increasingly questioned, among them health and environmental hazards. Also, Sokovic and Mijanovic (2001) noted that industries are being forced to develop less damaging cooling-lubrication strategies for machining processes.

Tawakoli *et al.* (2007) showed that one of the alternatives for reducing cutting fluid use is to optimize fluid flow, which occurs, for example, when MQL is used.

4.2.2 Minimum quantity lubrication

MQL is defined as a spray of a small amount of oil in a compressed air jet, directed to the cutting zone, minimizing the high flow rates of conventional flood coolant application (Obikawa *et al.*, 2006).

In MQL, the aerosol formed by the oil droplets carried by the compressed air provides efficient lubrication. Conventional cutting fluids, due to the use of additives and low application pressures, are not able to penetrate the cutting zone effectively, leaving MQL as more effective in this case.

A further advantage of MQL is that it makes cleaning easier, as chips, workpiece and tool are less impregnated with fluid after machining. In addition, during machining, as the workpiece is not completely flooded by fluid, as when using flood coolant, it is possible to visually monitor the process. In MQL, a low fluid flow is used, approximately from 10 to 100 mL/h, under a pressure of 4–6.5 bar (Attanasio *et al.*, 2006).

Although MQL promotes efficient lubrication, thereby reducing grinding-specific energies when compared with flood coolant application, under less severe machining conditions surface roughness values obtained are not satisfactory (Hafenbraedel and Malkin, 2001).

Brinksmeier *et al.* (1999) stated that a great challenge when using MQL concerns cooling, which limits its application to less severe machining conditions where a lower amount of heat is generated.

Choi *et al.* (2008) studied the application of cold compressed air to act as a coolant on cylindrical grinding of hardened steels. It was observed that the efficiency of air in reducing thermal damage was equivalent to that obtained with conventional fluids, for low depths of cut. However, residual stresses emerged, and the surface roughness values increased with the depth of cut. They inferred that this occurred due to the small amount of lubricant. Also, wheel wear increased as the grinding cycles progressed, but the forces and energies were lower than when using conventional cutting fluids.

Silva *et al.* (2007) reported that, when grinding SAE 4340 steel using MQL, surface roughness, diametric wheel wear, grinding forces and residual stresses were improved, due to improved sliding of abrasive grains on the contact zone.

Heisel *et al.* (1998) and Klocke *et al.* (2009) listed the advantages of MQL as compared with the conventional cutting fluids used in grinding:

- small amount of fluid used, avoiding the necessity of additional equipment for fluid circulation;
- no need for filtering or recycling processes;
- lower fluid maintenance costs;
- workpieces and machined chips are almost clean at the end of the process;
- minimized use of biocides and fungicides.

According to Machado and Diniz (2000), when compared with flood coolant lubrication, MQL requires investments to pressurize the air and to remove the machined chips. Productivity can also be lowered due to thermal damage of the machine tool components. A good exhaust system must be installed, so that the emissions can be safely removed from the environment. Also, the noise generated by the compressed air system must be minimized.

Tawakoli *et al.* (2009) state that, for efficient use of MQL, the effect of grinding parameters and workpiece materials must be taken into account. Also, satisfactory lubrication provides better grain slipping at the contact zone, which improves surface quality.

4.2.3 Thermal damage during grinding

The large amount of energy per unit of removed material volume is converted mainly to heat in the cutting zone, which can damage the surface integrity of a ground workpiece. Thermal damage to the material surface can affect it significantly, reducing wear and fatigue resistances, due to promotion of nucleation and propagation of cracks.

One of the main sources of damage to the workpiece surface is grinding burn. A visible burn can occur, characterized by an oxide layer on the surface. However, the absence of a visible layer does not indicate the complete absence of thermal damage. Such damage arises from successive heating (during material removal) and cooling (by the cutting fluid). These thermal cycles can induce microstructural transformations on the material surface, thus impairing the mechanical properties achieved with the previous processing (Silva *et al.*, 2007).

The increased efficiency of cooling-lubrication techniques can minimize the incidence of thermal damage (Malkin and Guo, 2007).

4.2.3.1 Visible burn (oxidation)

The oxidation characterized by an alteration of color is the most obvious and, most often, the first visible indication of thermal damage. Its color varies from light brown to dark blue, normally occuring above 450 °C; the lightest colors indicate oxidation at lower temperatures, but can also emerge at higher temperatures with lower cutting speeds (Marinescu *et al.*, 2004).

4.2.3.2 Hardness losses

Grinding temperatures may not be high enough to cause visible burn. However, in some situations, grinding overcomes the previous tempering temperature, resulting in surface hardness losses.

When metallographic analyses are conducted, the region affected by this retempering presents itself as a darker tone, and is usually known as a black layer. It does not indicate that microstructural alterations have occurred, but, when martensite suffers this super-tempering, there is a loss in hardness. The depth (in relation to the surface) where black layers can be found is no greater than 100 µm (Marinescu *et al.*, 2004).

4.2.3.3 Rapid quenching – white layer

In some situations during grinding, temperatures above the austenitizing temperature are achieved, and cooling at a fast rate can cause rapid quenching on the ground surface. Normally, the layer formed on the ground surface is thin, followed by a subsurface region with hardness losses (black layer). Due to its light color, when viewed under a microscope, it is usually called a white layer.

The white layer is detrimental to workpiece integrity, as the hardened surface becomes extremely brittle, and tension residual stresses arise, which drastically reduce wear and fatigue resistances.

4.2.4 Cubic boron nitride wheel

The superabrasive CBN was introduced in 1969 by General Electric, using the brand name Borazon (Salmon, 1992).

According to Guo *et al.* (2007), CBN wheels are largely used as alternatives to conventional alumina and silicon carbide wheels in many industrial sectors. As CBN possesses higher wear resistance, lower grinding temperatures can be achieved, reducing the likelihood of thermal damage and residual stresses.

Due also to the high thermal conductivity (1300 W/m.K), a small amount of heat is transferred to the workpiece (about 20 per cent), much lower than when alumina wheels are used (60–70 per cent), thus minimizing the incidence of thermal damage (Marinescu *et al.*, 2007).

CBN is a synthesized material obtained by the allotropic transformation of hexagonal boron nitride, under high pressures and temperatures. With extraordinary hardness, independent of crystal orientation, it possesses some advantages when compared with diamond, as it allows the machining of ferrous metals. Despite the greater hardness of diamond, its use in machining of steels is limited, as there is a tendency for carbon (existing in its crystal structure) to diffuse into the bulk of the workpiece, under typical machining pressures and temperatures.

A limiting factor for the application of superabrasive wheels is their high cost. However, when a global cost analysis is conducted, CBN is seen to provide lower grinding cycles, along with fewer tool changes and stops, related to wheel wear. All these factors can lead to a reduction of about 40 per cent of the total process costs (Fusse *et al.*, 2004).

4.3 Wheel cleaning

4.3.1 Wheel clogging

Wang *et al.* (2000) noted that, when the energy at the wheel/workpiece interface generates an increase in local temperature, machined chips are most likely to adhere to the abrasive wheel pores. Workpiece temperature therefore increases, causing thermal damage such as surface burn, thus reducing the surface quality and increasing wheel wear. Therefore, the lower the grinding-specific energy, the lower the probability of abrasive grains being loosened and of damage (Salmon, 1992).

Cameron *et al.* (2010) explained the phenomenon of wheel clogging as follows: when machined chips are not completely removed from the cutting zone by the cutting fluid, they lodge in the wheel pores, preventing perfect penetration of the cutting fluid in the cutting zone, and thus further hampering wheel cleaning. These lodged chips affect the efficiency and quality of the grinding operation, as they increase the contribution of elastic and plastic deformation to total grinding energy. Thus, the process energy will increase, as will the heat in the cutting zone.

When grinding takes place, chips are generated in and removed from the cutting zone; however, if MQL is used, these chips tend to mix with the MQL oil and lodge in the wheel pores (Figure 4.1). As the grinding process progresses, this 'grout' loads the wheel pores until there is contact between the lodged chips and the workpiece. Thus, the

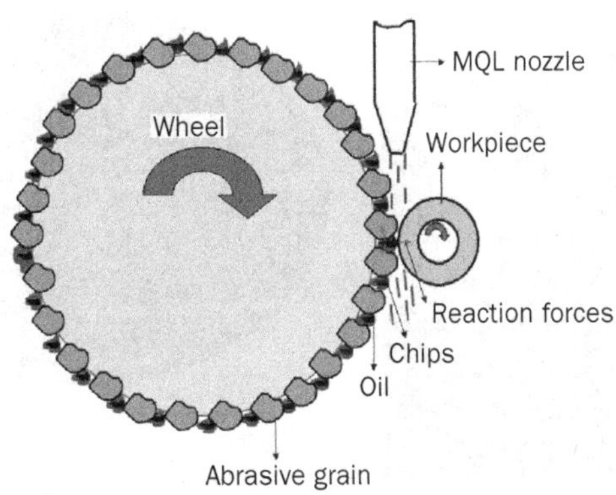

Figure 4.1 **Wheel clogging**

friction between the two components will cause deformation – elastic and plastic – only on the latter, which will increase the cutting power, surface roughness and roundness error of the workpiece, as well as wheel wear. Also, these lodged chips will generate reactions on the abrasive grains, causing increased wear due to grain loosening. Moreover, the workpiece will be scratched by the chips, worsening its final quality.

According to Sinot *et al.* (2006), some alloys are difficult to machine, e.g. titanium alloys, which clog the wheel pores more easily when metal particles are compressed and lodged between the grains. With high removal rates, the phenomenon is more intense, and some wheels are more prone to clogging than others.

Sinot *et al.* (2006) also describe two ways to avoid wheel clogging: to use an open structure wheel, which can increase the probability of bond fracture; and to re-dress the wheel more frequently, which increases process costs.

Figure 4.2 shows images with the grout in grey, clearly visible on the cutting surface, and a detailed view of a clogged wheel.

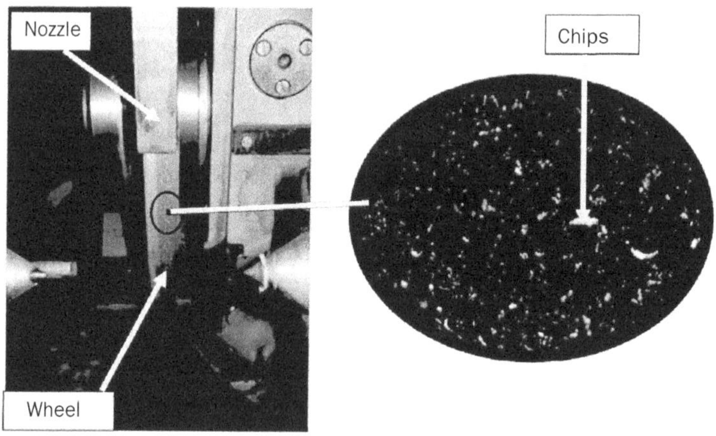

Figure 4.2 Grinding wheel with lodged chips (100×)

4.3.2 Wheel cleaning systems

4.3.2.1 Laser cleaning

Jackson *et al.* (2007) used a carbon dioxide laser to clean the wheel cutting surface. According to the same authors, a continuous or pulsing wave can be used. The laser passes through a convex lens and irradiates on the wheel surface, with a great amount of energy. This energy is capable of cutting the wheel, removing abrasive grains and lodged chips. Therefore, this irradiation must be controlled in order not to damage the tool, by means of two main parameters: laser beam energy flux, and the duration of the laser beam. By adjusting focal offset, laser irradiation energy flux can be controlled. Control of laser pulse duration and pulse frequency can also adjust the level of laser irradiation power.

4.3.2.2 Wheel cleaning with a secondary cutting fluid jet

Cameron *et al.* (2010) studied creep-feed grinding of steel with alumina wheels, using a secondary cutting fluid system (flood coolant, with the same cutting fluid as the main application) which applied a high-velocity (highest value of 143 m/s, corresponding to an internal pressure of 13.8 MPa) jet to clean the wheel surface, in addition to the normal flood coolant cooling-lubrication method. The stand-off distance of the cleaning nozzle was 12.7 mm. The distance ensured that the 30° spread of the cleaning jet would fully cover the workpiece width.

The addition of this cleaning system minimized grinding burn, when compared with the condition with only one cutting fluid flow. Cameron *et al.* (2010) also concluded that the chips lodged in the wheel pores increased the grinding forces, thus giving rise to burn.

They also verified that specific energy at 120 m/s was 33 per cent lower than when no cleaning was used. Thus, a higher efficiency was achieved, with lower tendency for thermal damage.

The system provided wheel cleaning, as well as cooling. The flow rate used for the jet was 17.8 L/min, and the incidence angles were: normal to the cutting surface, 26° and 90° (tangential). According to Cameron *et al.* (2010), when the nozzle was oriented normally to the wheel, the cutting fluid penetrated deeply; when aimed tangential to the wheel, it was reflected. Therefore, it was noted that a greater reduction in tangential forces was found when using an angle of incidence of 26°, because the cutting fluid could spread on the wheel surface, encompassing a broader area.

Webster (2008) described a situation in which an independent radial jet was added to clean the wheel, along with a tangential cooling jet. A synthetic water-based cutting fluid was used for flood coolant, in all nozzles. For the radial nozzle, the jet was applied at a flow rate of 64.3 L/min and a pressure of 413.7 kPa (60 p.s.i.). Webster reported that the optimized system provided better results, lowering wheel wear, reducing dressing and improving surface integrity.

4.3.2.3 Wheel cleaning with a compressed air jet

According to Lee *et al.* (2002), the compressed air jet provides an alternative to prevent wheel clogging, as it removes a larger proportion of the impurities that adhere to the tool. Thus, the lower pressure of the jet, the lower will be its cleaning efficiency.

They concluded that a greater depth of cut can be used, as wheel wear is lowered and also it is possible to obtain better surface quality, with narrow geometric and shape tolerances, when using compressed air jets to clean the wheel cutting surface.

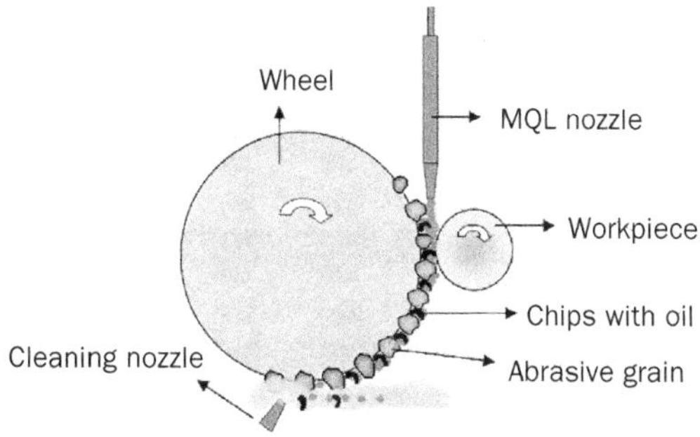

Figure 4.3 **Effect of compressed air jet on wheel cleaning**

It was expected that, when using MQL, wheel cleaning would also occur, similar to that verified by Lee *et al.* (2002), as shown in Figure 4.3.

Oliveira *et al.* (2012) recently reported the use of a compressed air jet for wheel cleaning, when using MQL in CBN grinding. The results showed that an incidence angle of 30° is more suitable for penetration in the cutting zone, providing efficient cleaning and better results in almost all the process outputs analysed, such as surface roughness, roundness errors, wheel wear and acoustic emission signals, when compared with traditional MQL (without cleaning).

Reducing fluid flow, as when using MQL, associated with better results of the output variables, indicates that this improvement of the cooling-lubrication conditions in grinding operations may reduce environmental and health hazards, contributing to a cleaner, faster and more cost-effective manufacturing process.

4.3.3 MQL with water addition

Even with the development of MQL, the search for new cutting fluid techniques has steadily progressed. Studies concerned with reduction of cutting fluid usage in the machining process, as well as minimization of their drawbacks, have continued. As a consequence of these efforts, the addition of water in MQL fluid has recently appeared as an improvement.

According to Yoshimura *et al.* (2005), this modification provides extreme cooling capability, as water droplets tend to carry a thin coating of oil. When these droplets impact on the workpiece surface, water rapidly evaporates due to the small size of the droplets, cooling the surface owing to their sensible and latent heats, thus leaving the oil to act as a lubricant.

This high cooling capacity is of great importance to dimensional and geometric tolerances, to surface roughness and to other phenomena involving the cutting tool and machined surface.

Yoshimura *et al.* (2005) stated that the addition of water in the MQL technique improves lubrication when compared with traditional MQL, reducing the friction between wheel and workpiece. Also, their results show that MQL with water can reduce tool wear, axial cutting forces and tool contact width, while providing more curled chips (as a consequence of cooling).

Yoshimura *et al.* (2005) concluded that MQL with water is more efficient than traditional MQL. Besides the reduction in cutting forces and tool wear, it also promotes sufficient lubrication to suppress local temperature rise in the contact zone, as well as sufficient cooling capacity due to water evaporation, minimizing increases in temperature.

It must be noted, however, that the results and conclusions presented by the above researchers refer to machining processes which use cutting tools with defined geometry. Few studies have been conducted concerning this new technique, and no results have been published concerning MQL with water additions in grinding.

4.4 Conclusions

Optimization of the MQL technique in grinding has been widely researched to minimize the occurrence of wheel clogging (wheel cleaning) and the use of cutting fluids (MQL with addition of water).

However, the results and conclusions presented by the authors discussed here refer to specific machining conditions, which may be not reliable for different grinding parameters, wheels and workpiece materials. Therefore, much work remains to be done to obtain optimal conditions for MQL in grinding, which can provide results similar or better to flood coolant application.

The increasing need for sustainable manufacture should stimulate further research and studies of alternative cooling-lubrication methods, as well as improvements in existing techniques. Along with workpiece quality results, cost estimations should be done for each case to ensure the new method (or improvement) is feasible in terms of industrial application.

References

Alves J A C, Fernandes U B, Diniz A E, Bianchi E C, Aguiar P R, Canarim R C (2009), 'Analysis of the influence of sparkout time on grinding using several lubrication/cooling methods', *Journal of the Brazilian Society of Mechanical Sciences and Engineering*, **31**(1): 47–51.

Alves M C S (2007), '*Influência dos métodos de lubri-refrigeração na retificação do aço ABNT 4340 usando rebolo de CBN*', PhD Thesis, UNESP – Bauru.

Alves M C S, Bianchi E C, Aguiar P R, Canarim R C (2011), 'Influence of optimized lubrication-cooling and minimum quantity lubrication on the cutting forces, on the geometric quality of the surfaces and on the micro-structural integrity of hardened steel parts', *Matéria*, **16**(3): 754–66.

Attanasio A, Gelfi M, Giardini C, Remino C (2006), 'Minimal quantity lubrication in turning: effect on tool wear', *Wear*, **260**(3): 333–8.

Bartz W J (1998), 'Lubricants and the environment', *Tribology International*, **31**(1–3): 35–47.

Brinksmeier E, Heinzel C, Wittman M (1999), 'Friction, cooling and lubrication in grinding', *CIRP Annals – Manufacturing Technology*, **48**(2): 581–98.

Byrne G, Scholta E (1993), 'Environmentally clean machining processes—a strategic approach', *CIRP Annals–Manufacturing Technology*, **42**(1): 471–4.

Cameron A, Bauer R, Warkentin A (2010), 'An investigation of the effects of wheel cleaning parameters in creep-feed grinding', *International Journal of Machine Tools and Manufacture*, **50**(1): 126–30.

Catai R E, Silva L R, Bianchi E C, Aguiar P R, Zilio F M, Valarelli I D, Salgado M H (2008), 'Performance of aerodynamic baffles in cylindrical grinding analyzed on the basis of air layer pressure and speed', *Journal of the Brazilian Society of Mechanical Sciences and Engineering*, **30**(1): 47–50.

Choi T J, Subrahmanya N, Li H, Shin Y C (2008), 'Generalized practical models of cylindrical plunge grinding processes', *International Journal of Machine Tools and Manufacture*, **48**(1): 61–72.

Dhar N, Islam S, Kamruzzaman M (2006), 'Effect of minimum quantity lubrication (MQL) on tool wear, surface roughness in turning AISI-4340 steel', *Journal of Materials Processing Technology*, **172**(2): 299–304.

Diniz A E, Marcondes F C, Coppini N L (2008), *Tecnologia da usinagem dos materiais*. Artiliber Editora Ltd.

Ebbrell S, Woolley N H, Tridimas Y D, Allanson D R, Rowe W B (2000), 'Effects of cutting fluid application methods on the grinding process', *International Journal of Machine Tools and Manufacture*, **40**(2): 209–23.

Fratila D, Caizar C (2011), 'Application of Taguchi method to selection of optimal lubrication and cutting conditions in face milling of $AlMg_3$', *Journal of Cleaner Production*, 19(6–7): 640–5.

Fusse R Y, França T V, Catai R E, Silva L R, Aguiar P R, Bianchi E C (2004), 'Analysis of the cutting fluid influence on the deep grinding process with a CBN grinding wheel', *Materials Research*, 7(3): 451–7.

Guo C, Shi Z, Attia H, McIntosh D (2007), 'Power and wheel wear for grinding nickel alloy with plated CBN wheels', *CIRP Annals – Manufacturing Technology*, 56(1): 343–6.

Hafenbraedl D, Malkin S (2001), 'Tecnologia ambientalmente correta para retificação cilíndrica interna', *Machines and Metals Magazine*, 37(426): 40–55.

Heisel U, Lutz D, Wassmer R, Walter U (1998), 'The minimum quantity lubricant technique and its application in the cutting process'. *Machines and Metals Magazine*, 386: 22–38.

Hong S Y, Broomer M (2000), 'Economical and ecological cryogenic machining of AISI 304 austenitic stainless steel', *Clean Technologies and Environmental Policy*, 2(3): 157–66.

Howell J R, Lucke W E, White E M (2006), *Health and Safety Aspects in the Use of Metalworking Fluids*. London: Taylor & Francis Group.

Irani, R A, Bauer R J, Warkentin A (2005), 'A review of cutting fluid application in the grinding process', *International Journal of Machine Tools and Manufacture*, 45(15): 1696–705.

Jackson M J, Khangar A, Chenc X, Robinson G M, Venkatesh V C, Dahotre N B (2007), 'Laser cleaning and dressing of vitrified grinding wheels', *Journal of Materials Processing Technology*, 185(1–3): 17–23.

Klocke F (2009), *Manufacturing Processes 2: Grinding, Honing, Lapping*. Berlin: Springer.

Lee S W, Le Y C, Jeond H D, Choi H Z (2002), 'The effect of high pressure air jet on form accuracy in slot grinding', *Journal of Materials Processing Technology*, 128(1–3): 67–72.

Machado A R, Diniz A E (2000), 'Advantages and disadvantages of the use of the cutting fluids'. Machining Congress 2000, São Paulo, SP, Brazil.

Malkin S, Guo C (2007), 'Thermal analysis of grinding', *CIRP Annals – Manufacturing Technology*, 56(2): 760–82.

Marinescu I D, Hitchner M, Uhlmann E, Rowe W B, Inasaki I (2007), *Handbook of Machining with Grinding Wheels*. CRC Press/Taylor & Francis Group.

Marinescu I D, Rowe W B, Dimitrov B, Inasaki I (2004), *Tribology of Abrasive Machining Processes*. William Andrew Inc.

Min S, Inasaki I, Fujimura S, Wada T, Suda S, Wakabayashi T (2005), 'A study on tribology in minimal quantity lubrication cutting', *CIRP Annals – Manufacturing Technology*, **54**(**1**): 105–8.

Obikawa T, Kamata Y, Shinozuka J (2006), 'High-speed grooving with applying MQL', *International Journal of Machine Tools and Manufacture*, **46**(**14**): 1854–61.

Oliveira D J, Guermandi L G, Bianchi E C, Diniz A E, Aguiar P R, Canarim, R C (2012), 'Improving minimum quantity lubrication in CBN grinding using compressed air wheel cleaning', *Journal of Materials Processing Technology*, **212**(**12**): 2559–68.

Pawlak Z, Klamecki B E, Rauckyte T, Shpenkov G P, Kopkowski A (2004), 'The tribochemical and micellar aspects of cutting fluids', *Tribology International*, **38**(**1**): 1–4.

Ren Y H, Zhang B, Zhou Z X (2009), 'Specific energy in grinding of tungsten carbides of various grain sizes', *CIRP Annals – Manufacturing Technology*, **58**(**1**): 299–302.

Sadeghi M H, Hadad M J, Tawakoli T, Emami M (2009), 'Minimal quantity lubrication-MQL in grinding of Ti–6Al–4V titanium alloy', *International Journal of Advanced Manufacturing Technology*, **44**(**5**): 487–500.

Salmon S C (1992), *Modern Grinding Process Technology*. New York: McGraw-Hill.

Silva L R, Bianchi E C, Fusse R Y, Catai R E, França T V, Aguiar P R (2007), 'Analysis of surface integrity for minimum quantity lubricant – MQL in grinding', *International Journal of Machine Tools and Manufacture*, **47**(**2**): 412–18.

Sinot O, Chevrier P, Padilla P (2006), 'Experimental simulation of the efficiency of high speed grinding wheel cleaning', *International Journal of Machine Tools and Manufacture*, **46**(**2**): 170–5.

Sokovic M, Mijanovic K (2001), 'Ecological aspects of the cutting fluids and its influence on quantifiable parameters of the cutting processes', *Journal of Materials Processing Technology*, 109(1–2): 181–9.

Stanford M, Lister P M, Kibble K A, Morgan C (2007), 'Investigation into the use of gaseous and liquid nitrogen as a cutting fluid when turning BS 970-80A15 (En32b) plain carbon steel using WC-Co uncoated tooling', *Journal of Materials Processing Technology*, 209(2): 961–72.

Tawakoli T, Westkaemper E, Rabiey W (2007), 'Dry grinding by special conditioning', *International Journal of Advanced Manufacturing Technology*, 33(3–4): 419–24.

Tawakoli T, Hadad M J, Sadeghi M H, Daneshi A, Stöckert S, Rasifard A (2009), 'An experimental investigation of the effects of workpiece and grinding parameters on minimum quantity lubrication—MQL grinding International', *Journal of Machine Tools and Manufacture*, 49(12–13): 924–32.

Tawakoli T, Hadad M J, Sadeghi M H (2010), 'Influence of oil mist parameters on minimum quantity lubrication–MQL grinding process', *International Journal of Machine Tools and Manufacture*, 50(6): 521–31.

Wang C M, Cargill G S, Chan H M, Harmer M P (200), 'Structure of Y and Zr segregated grain boundaries in alumina', *Interface Science*, 8(2–3): 243–55.

Webster J A (2008), 'Coolant calculus: directing coolant into the right place at the right speed, in the right quantity', *Cutting Tool Technology*, 60(2): 58–66.

Weinert K, Inasaki I, Sutherland J W, Wakabayashi T (2004), 'Dry machining and minimum quantity lubrication', *CIRP Annals – Manufacturing Technology*, 53(2): 511–37.

Yoshimura H, Itoigawa F, Nakamura T, Niwa K (2005), 'Development of nozzle system for oil-on-water droplet metalworking fluid and its application on practical production line', *JSME International Journal Series C*, 48(4): 723–29.

5

Electrical discharge machining: study on machining characteristics of WC/Co composites

K. Palanikumar[1] *and J. Paulo Davim*[2]

Abstract: Electrical discharge machining (EDM) is a non-traditional machining process based on removing material from a part by means of a series of repeated electrical discharges between tools, called electrodes, and the part being machined in the presence of a dielectric fluid. EDM is currently widely used in industry for high-precision machining of all types of conductive material such as metals, metallic alloys, graphite and even some ceramic materials, regardless of hardness. In spite of their exceptional mechanical and chemical properties, ceramic materials have only been partially accepted in the field of industrial applications, owing to difficulties with processing and the high cost associated with their manufacture. Over the past few years, advances in the field of EDM make the application of this technology available for the manufacture of conductive ceramic materials. This chapter first presents an overview of the general classification, manufacturing and various applications of EDM

[1] Department of Mechanical Engineering, Sri Sai Ram Institute of Technology, Sai Leo Nagar, Chennai 600 044, India. E-mail: *palanikumar_k@yahoo.com; palanikumar@sairamit.edu.in*

[2] Department of Mechanical Engineering, University of Aveiro, Campus Santiago, 3810-193 Aveiro, Portugal. E-mail: *pdavim@ua.pt*

Published by Woodhead Publishing Limited, 2013

and secondly presents the application of EDM for the machining of tungsten carbide/cobalt (WC/Co)-cemented carbide, an important composite material that is used in the manufacture of cutting tools, dies and other special tools. It has high hardness and excellent resistance to shock and wear, but is difficult to machine easily using conventional techniques. Experiments are performed on a newly designed experimental set-up developed in the laboratory. The response surface methodology is used to identify the most influential parameters for maximizing metal removal rate and for minimizing surface roughness.

Key words: Analysis of variance, die sinking EDM, electrical discharge machining (EDM), non-traditional machining, response surface methodology, tungsten carbide/cobalt, WC/Co composites.

5.1 Introduction

Electrical discharge machining (EDM) is one of the most important non-conventional machining processes and has recently become increasingly popular. In EDM the material is removed from a part by a series of repeated electrical discharges between the tool, the 'electrode', and the part being machined in the presence of a non-conductive fluid, the 'dielectric fluid'. The dielectric fluid provides a means of flushing and is pumped through the arc gap between electrode and workpiece. This process removes suspended particles of the workpiece and the electrode from the work area. EDM is widely used in industry for high-precision machining of all types of conductive materials such as metals, metallic alloys, graphite and even some ceramic materials, regardless of hardness. EDM is normally used for machining extremely hard materials to form complex shapes. EDM is mainly applied in the die and mould-making industry and also in the construction of prototypes. The introduction of numerically controlled equipment enabled various new EDM technologies such as deep sinking EDM, contouring EDM, wire EDM and milling EDM (Lauwers *et al.*, 2012). EDM has proved to be a valuable operation in the machining of super-tough electrically conductive materials such as new space-age materials, which are difficult to machine using conventional methods. EDM is mainly used in machining of intricate shapes that would be impossible to produce with conventional cutting tools. This machining process is continuously finding further applications in the machining industry (Luis *et al.*, 2005).

Tungsten carbide/cobalt (WC/Co) composite is one of the most important classes of materials used for producing cutting tools, dies and other special tools and components having high hardness, strength and wear resistance over a wide range of temperatures. It has high specific strength and cannot be processed easily by conventional machining techniques (Lee and Li, 2001). It is possible to machine this material with some conventional methods, but high accuracy is required for machining complex shapes that cannot be achieved by conventional machining. In general, grinding is the only accepted conventional method for machining various kinds of cemented carbides, especially when high surface quality is required. However, during the grinding process, local heat generation due to machining causes microcracks on the workpiece material (Kanagarajan *et al.*, 2009). This is especially true in the machining of WC/Co composites, because WC and Co are two quite different phases. As a result, a complementary finishing operation might be necessary to remove these cracks and to produce a good surface quality when conventional machining is used (Llanes *et al.*, 2004; Puertas *et al.*, 2004). Among non-conventional machining methods, electrochemical and EDM are suitable for machining of WC/Co composites. However, in electrochemical machining, the stock removal rate is very low because of the creation of a resistant oxide layer on the surface of the workpiece. Therefore, EDM is usually used for machining these materials. Given its ability to machine WC/Co composites, the EDM process is very important in machining cemented carbide materials, and many studies have been done in this regard (Zhang *et al.*, 1997; Lee and Li, 2001; Mahdavinejad and Mahdavinejad, 2005). Electro-discharge machine manufacturers and users are clearly interested in acquiring better stability and higher productivity in the machining process. The higher rate of material removal with desired accuracy and minimal surface damage make the EDM operation less costly, and hence the process is more economically viable and affordable. However, due to a great number of variables and a variety of products, optimal machining performance is rarely achieved (Chen and Mahdavian, 1999; Kanagarajan *et al.*, 2008a,b). It is necessary to investigate how the erosion parameters affect the machining process. The results will provide significant information to achieve optimal process performance (Chen and Mahdavian, 1999). Lajis *et al.* (2009) used the Taguchi method to analyse the effect of machining parameters on the machining performance of EDM parameters such as peak current, voltage, pulse duration and interval time. They found that peak current significantly affects the electrode wear ratio and surface roughness, while the pulse duration mainly affects the material removal rate (MRR). A similar study by Tomadi *et al.* (2009) reported the influence of operating parameters on machining characteristics such as surface quality, MRR and electrode wear rate during the EDM of tungsten carbide.

George *et al.* (2004) have conducted experiments and analysed the results using statistical analysis of variance (ANOVA) to identify the process parameters which have an influence on the machining. Lee and Li (2001) studied the effect of machining parameters on the machining characteristics in EDM of tungsten carbide. The results indicated that better performance was achieved with the electrode as cathode. The workpiece as anode and tools with negative polarity gave better MRR, lower tool wear and better surface finish. Lin and Lin (2002) have developed mathematical models for EDM characteristics, based on the phenomenon of crater formation due to a single spark. Much effort has been made to forecast the profile and also to find the volume of the crater, the MRR and the tool wear rate. A comparative study on the performance of three different types of electrodes for the EDM of tungsten carbide was conducted by Janmanee and Muttamara (2010). The electrode materials studied were graphite (Poco EDM-3), copper–graphite (Poco EDM-C3) and copper–tungsten (solid). The performance of the electrodes was evaluated by studying the effect of discharge current, on-time, off-time, open-circuit voltage and electrode polarity on the MRR for machining tungsten carbide. Chen and Hsu (2003) have studied electrical discharge machining of non-contact seal face grooves in WC and other materials. They found that WC has better surface finishes by using negative-polarity. In a recent study, Sano *et al.* (2007) have machined tungsten carbide with poly-crystalline diamond (PCD), comparing the EDM performance of the PCD tool with that of commonly used EDM electrodes: copper (Cu) and copper tungsten (CuW). The results indicated that PCD electrodes perform better in machining of WC-Co.

Review of the above studies indicates that modeling of the EDM process can be regarded as a problem of correlating the input parameters of the process with its output parameters. This can be dealt with by means of response surface methodology (RSM), which is an empirical modeling approach for determining the relationship between various process parameters and responses with the various desired criteria and assessing the significance of these process parameters on the coupled responses (Kanagarajan *et al.*, 2008b). It represents a sequential experimentation strategy for building and optimizing the empirical model. In the present paper, an attempt has been made to obtain optimal settings for the process parameters, which may yield optimum MRR and minimum surface finish (R_a) on the components processed by EDM. RSM has been used to plan and analyse the experiments (Puertas and Luis, 2004).

To understand the efficient use of EDM for ceramic materials, this chapter first presents an overview of the general classification, manufacturing and various applications of the EDM process and secondly focuses on the die-sinking EDM of a ceramic material such as cobalt-bonded tungsten carbide (WC/Co) and its field of applications. Analysis was made of the influence of electrode rotation (S), pulse on-time (T), current (A) and flushing pressure (P) on the output responses MRR and surface roughness. The experiments are based on the methodology of Box and Bhenken (1960). Mathematical relations are established between the parameters and responses MRR and surface roughness. The effect of the parameters and their interactions are presented in detail and discussed.

5.2 Principles of the EDM process

EDM is a thermoelectric process used to remove metal via a series of discrete sparks between the metal and workpiece. In EDM an electric spark is used as the cutting tool to cut (erode) the workpiece and produce the finished part to the required shape. Electricity flows through the electrode in the form of a square wave attacking the points of least resistance on the workpiece, similar to a bolt of lightning hitting a tree before it hits the ground (http://www.polecraft.com.) The basic principle is shown in Figure 5.1.

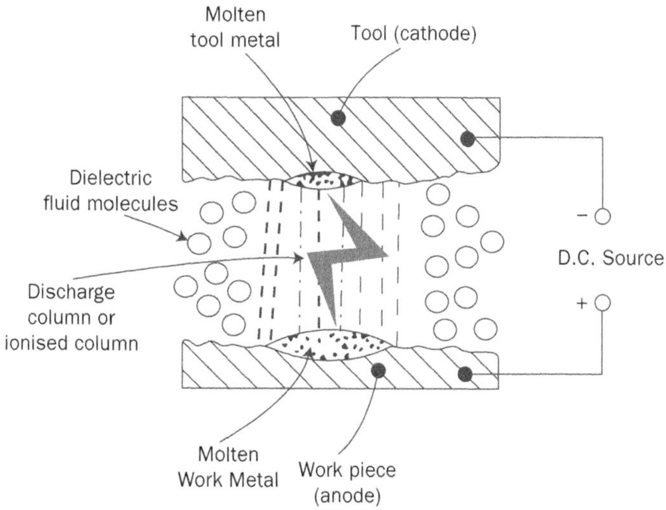

Figure 5.1 **Principle of the electric discharge machining process**

In EDM, material removal is performed by applying a pulsating (ON/OFF) high-frequency current through the electrode to the workpiece. This removes (erodes) very tiny pieces of material from the workpiece at a controlled rate. The electrode and the workpiece are both immersed in a liquid called dielectric fluid. The workpiece is normally the anode and the tool is the cathode. A thin gap of about 0.025 mm is maintained between the tool and the workpiece by a servo system. A suitably shaped tool with feed mechanism is used to get the proper shape and size of the work. The dielectric used in the system is continuously circulated and filtered. No grinding or finishing is required after the EDM operation. A schematic diagram of the EDM process is presented in Figure 5.2.

EDM is basically of three types according to the electrode used, namely die sinking EDM, wire-cut EDM and micro EDM mill. Die sinking EDM consists of an electrode and the workpiece

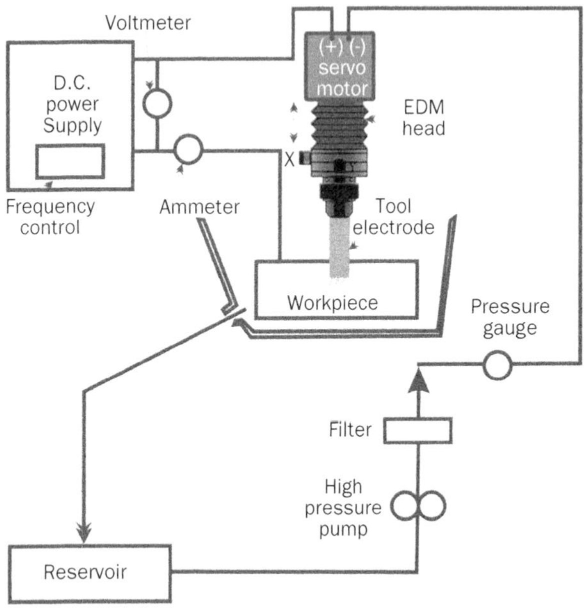

Figure 5.2 **Schematic of an electric discharge machine**

is submerged in an insulating liquid. The electrode and the workpiece are connected to a power supply in which the electrode is made according to the required shape. The machine uses the shape of the electrode to erode the inverse shape in the workpiece. This is one of the important accurate processes normally used for making mould cavities. The die sinking EDM arrangement is presented in Figure 5.3. In this form of EDM, the electrode is configured like the cavity desired in the workpiece. The machining speed depends upon the type of material, area of the material to be machined and the machining condition.

In wire EDM the electrode is a wire and is fed through an upper and lower diamond guide. The wire diameters usually range from 0.05 to 3.0 mm. The electrode is used only once and is discarded after it is used. Wire control is carried out by computer numerical control, which programs the path of

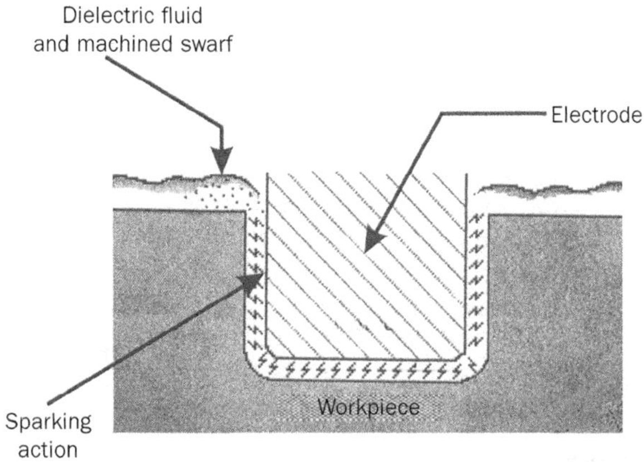

Figure 5.3 **Conventional die sinking EDM**

Source: www.materion.com

the wire. This form of EDM is the most accurate and can be used for both rough and finish machining. A schematic of travelling wire EDM is presented in Figure 5.4. A third type of EDM recently developed is micro EDM mill in which the electrode has diameters of 50 µm to 10 mm. The complex electrode is replaced by a set of moving standardized electrodes. The process is similar to the conventional die sinking EDM. Micro EDM complements wire EDM by providing the ability to drill pilot holes through heat-treated materials and carbide. It can be used for machining deep cavities up to 10 mm with nearly vertical walls of 1.5° taper, and also for machining of very thin (<0.1 mm) ribs. This technology is an important application in micro machining.

The EDM process is used for making different kinds of workpieces. The EDM machine is programmed for vertical machining, orbital, vectorial, polygon, spherical, circular, directional, helical, conical, spin and indexing machining cycles. These different machining cycles are used for making parts, and are presented in Figure 5.5.

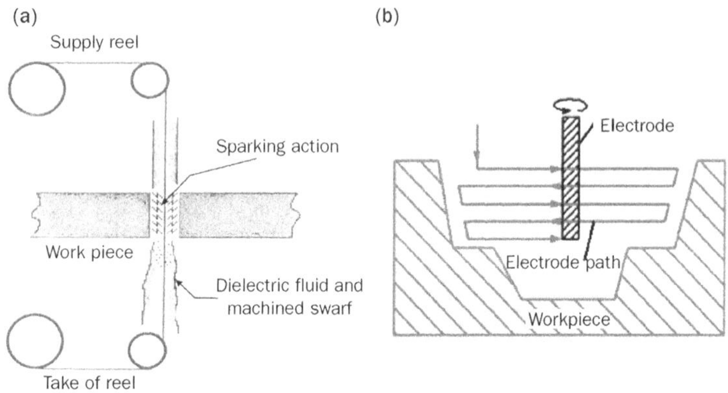

Figure 5.4 (a) Wire EDM, (b) micro EDM mill

Source: www.materion.com

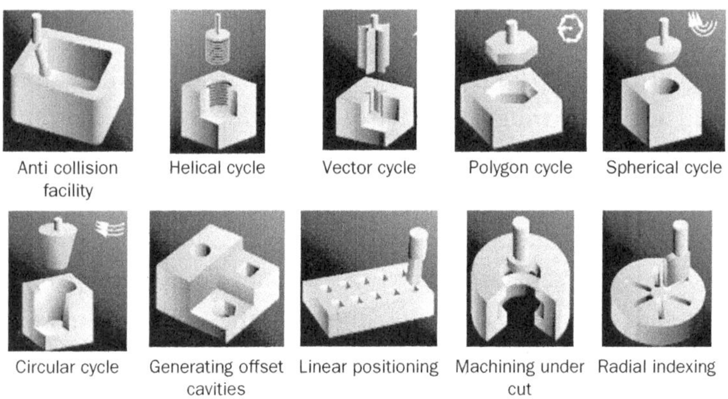

Figure 5.5 Tools of different cross sections and produced workpiece

Courtesy: Electronica Machine Tools Ltd, India

EDM has many advantages over conventional machining. Using EDM, electrically conductive material can be cut and has been used for machining hardened and heat-treated workpieces. Complex and intricate profiles can cut accurately, faster and at lower costs. Thin, fragile sections have been

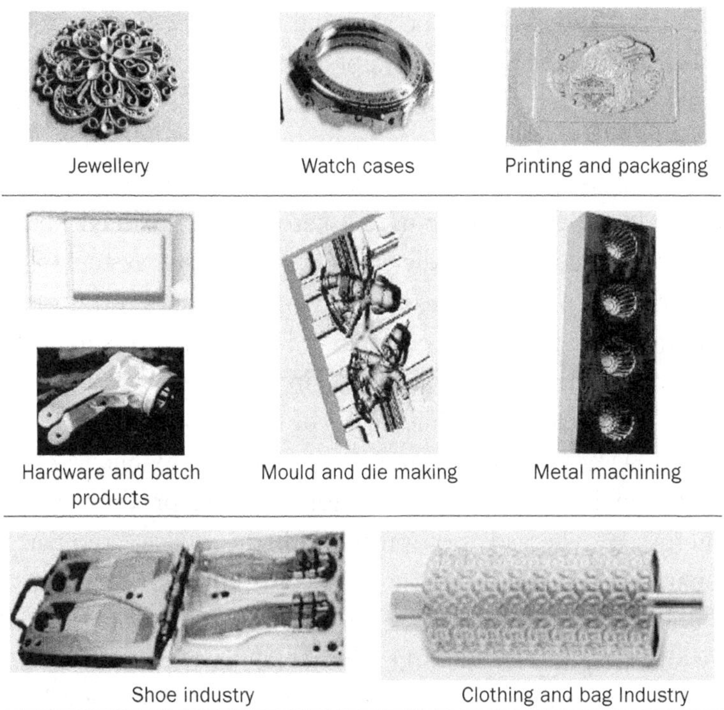

Figure 5.6 **Some applications of EDM machining**

Courtesy: Electronica Machine Tools Ltd, India

made easily and without burrs. EDM is mainly used in the tool and die industries. Today, EDM is an integral part of the process for making prototypes and production parts. Some of the applications of EDM are presented in Figure 5.6.

5.3 Machining experiments

Analysing the machining quality for WC-Co composites is an important concern. Machining experiments were thus carried out on an EDM die-sinking machine (M100-model; Electronica

Industries Ltd, India). The machine is equipped with proper transistor-switched power supply. The electrode used for the machining operation is made of copper. The electrode is fed downwards under servo control into the workpiece. The electrode used is cylindrical with a diameter of 12 mm and the dielectric fluid system used is kerosene jet flushing. This is used to circulate the dielectric fluid in the system. The experiments are performed by varying the pulse on-time, tool rotational speed, peak current and flushing pressure. The response parameters considered in the present study are MRR and surface roughness (R_a) of the workpiece.

In EDM, MRR is an important aspect. A higher MRR is preferred because it is proportional to the product of the energy transferred per pulse and the pulse frequency. Changing the pulse current at a constant frequency varies the energy of the pulse. In each test, the MRR is calculated by the weight loss method. The MRR is expressed as the workpiece removal weight (WRW) under a period of machine time (T) in minutes:

$$\text{MRR (mg/min)} = \frac{\text{Volume of material removed from the part}}{\text{Time of machining}} = \frac{\text{WRW}}{T} \quad [5.1]$$

In addition to MRR, EDM aims to produce a good part with low surface roughness. Therefore, surface roughness is another important index in evaluating the cutting performance (Benardos and Vosniakos, 2003).

Here, average surface roughness (R_a), as generally used in industry, is considered. Average surface roughness (Figure 5.7) is the integral of the absolute value of the roughness profile height over the evaluation length and is calculated as follows (Francis *et al.*, 2002; Rogelio *et al.*, 2003):

$$R_a = \frac{1}{L}\int_0^L \left[Y(x)\right] dx \quad [5.2]$$

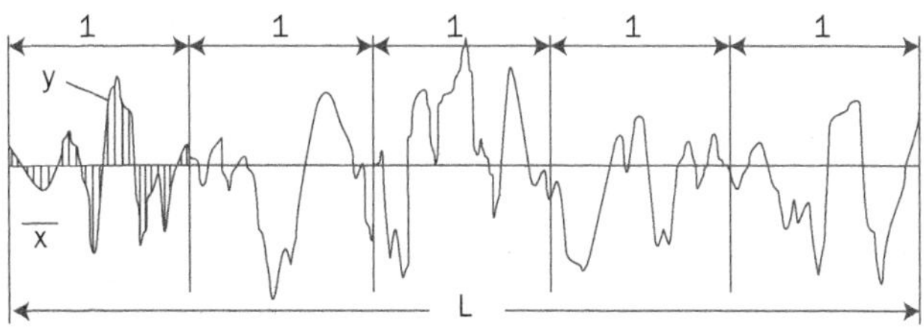

Figure 5.7 Surface roughness parameter (R_a)

where L is the length taken for observation and Y is the ordinate of the profile curve. Surface roughness was measured by using a surface roughness tester with 0.8 mm as the cut-off value.

A structured, organized method was used to determine the relationship between factors that affect the process and the output of that process, with a minimum number of experiments. A Box–Behnken design was used to conduct the experiments, similar to the central composite design. The Box–Behnken design avoids corner points and star points. A four-factor, three-level Box–Behnken design (Box and Behnken, 1960) is used. This design is suitable for exploration of quadratic response surfaces and constructs a second-order polynomial model. This helps in optimizing a process using a small number of experimental runs. The design consists of replicated centre points and the set of points lying at the midpoints of each edge of the multidimensional cube that defines the region of interest (Anees and Manzoor, 1996). Table 5.1 shows the factors and their levels in coded and actual values. For four variables, the design requires 27 experiments. The design is generated and analysed using the MINITAB statistical package.

The standard order of the experiment, its run order, the variables considered for the experiments (pulse on-time, tool rotational speed, peak current and flushing pressure), their setting levels and the responses measured (MRR and R_a) are presented in Table 5.2. The experiments are conducted

Table 5.1 Process parameters and their levels used for the investigation

Parameter	Symbol	Levels		
		-1	0	+1
Pulse on-time (μs)	*T*	200	500	1000
Electrode rotation (r.p.m.)	*S*	100	150	250
Peak current (A)	*C*	10	15	20
Flushing pressure (kg/cm^2)	*P*	1.0	1.5	2.0

Table 5.2 Experimental results

No.	Std. order	Run order	Pulse on-time (μs, *T*)	Electrode rotation (r.p.m., *S*)	Peak current (A, *C*)	Flushing pressure (kg/cm^2, *P*)	MRR (mg/min)	R_a (μm)
1.	17	1	200	150	15	1	23.54	4.95
2.	11	2	200	150	20	1.5	52.13	3.79
3.	22	3	500	250	10	1.5	9.66	3.8
4.	23	4	500	100	20	1.5	26.99	4.33
5.	2	5	1000	100	15	1.5	12.12	5.65
6.	26	6	500	150	15	1.5	20.45	4.13
7.	8	7	500	150	20	2	45.66	3.12
8.	14	8	500	250	15	1	32.41	4.63
9.	16	9	500	250	15	2	38.57	3.22
10.	10	10	1000	150	10	1.5	15.25	4.43
11.	27	11	500	150	15	1.5	22.12	4.08
12.	3	12	200	250	15	1.5	50.54	3.85
13.	9	13	200	150	10	1.5	28.09	4.14
14.	1	14	200	100	15	1.5	31.41	4.27
15.	18	15	1000	150	15	1	12.32	5.25
16.	20	16	1000	150	15	2	17.44	4.7
17.	7	17	500	150	10	2	19.86	4.05
18.	25	18	500	150	15	1.5	28.52	3.86
19.	24	19	500	250	20	1.5	45.51	4.23
20.	4	20	1000	250	15	1.5	10.58	4.74
21.	12	21	1000	150	20	1.5	14.89	5.13
22.	13	22	500	100	15	1	15.2	4.81
23.	15	23	500	100	15	2	26.52	4.28
24.	19	24	200	150	15	2	46.4	3.71
25.	21	25	500	100	10	1.5	10.95	4.47
26.	6	26	500	150	20	1	28.89	4.87
27.	5	27	500	150	10	1	17.62	4.63

Published by Woodhead Publishing Limited, 2013

based on preliminary experiments and our experience of the field. In previous studies (Kanagarajan *et al.*, 2008a,b) we conducted experiments using a central composite design. In the present study, the experiments are carried out using the Box–Behnken design in the machining of composites with fewer number of experiments and limited range of parameters. The experimental results are presented in Table 5.2.

5.4 Response surface modelling using Box-Behnken design

MRR and the surface finish are important in machining of WC-Co composites. While machining, high quality of the parts can be achieved only through proper machining conditions. To determine MRR and surface quality, it is necessary to employ theoretical models to predict the function of machining conditions (Sahin and Motorcu, 2008). RSM is one of the important techniques for determining and representing the cause-and-effect relationship between true mean responses and input control variables, influencing the responses as a two- or three-dimensional hyper surface (Godfrey and Kumar, 2006). Many researchers have applied RSM for modeling and analysis of the process parameters in manufacturing. For example, Palanikumar (2008) has used RSM for modeling the machining parameters for surface roughness in machining of glass fibre-reinforced plastic composites. RSM is simple and easy to use. In the present study, the experiments are conducted using a Box–Behnken experimental design technique and the empirical relations are established for correlating the interactive and higher-order influences of various machining parameters on MRR and surface roughness in machining of WC/Co composites.

Box–Behnken designs were devised by George E. P. Box and Donald Behnken in 1960 to achieve the following goals (http://en.wikipedia.org/wiki/Box%E2%80%93Behnken_design):

- Each factor, or independent variable, is placed at one of three equally spaced values. For conducting experiments at least three levels are required.
- The design can be sufficient to fit a quadratic model, which includes square effects and interaction effects between factors.
- The ratio of the number of experimental points to the number of coefficients in the quadratic model should be reasonable.

In many engineering fields, there is a relationship between an output variable of interest 'y' and a set of controllable variables $\{x_1, x_2, ..., x_n\}$. If the nature of the relationship between 'y' and 'x' values is known, then a model can be written in the form:

$$y = f(x_1, x_2, ... x_n) + \varepsilon \quad [5.3]$$

where ε represents the error associated with the response 'y'. If we denote the expected response by $E(y) = f(x_1, x_2, ... x_n) + \eta$, then the surface represented by:

$$\eta = f(x_1, x_2, ... x_n) \quad [5.4]$$

is called the response surface. Usually, a second-order model is utilized in response surface methodology (Montgomery, 1991):

$$y = \beta_0 + \sum_{i=1}^{k} \beta_i x_i + \sum_{i=1}^{k} \beta_{ii} x_i^2 + \sum_{i} \sum_{j} \beta_{ij} x_i x_j + \varepsilon \quad [5.5]$$

The β coefficients are calculated using the least-squares method. The response surface methodology is used to find values of the controllable parameters that result in optimization of the response or to discover what values of x will result in a product or process satisfying several requirements or specifications (Kwak, 2005). A flow chart of the steps used in RSM is presented in Figure 5.8.

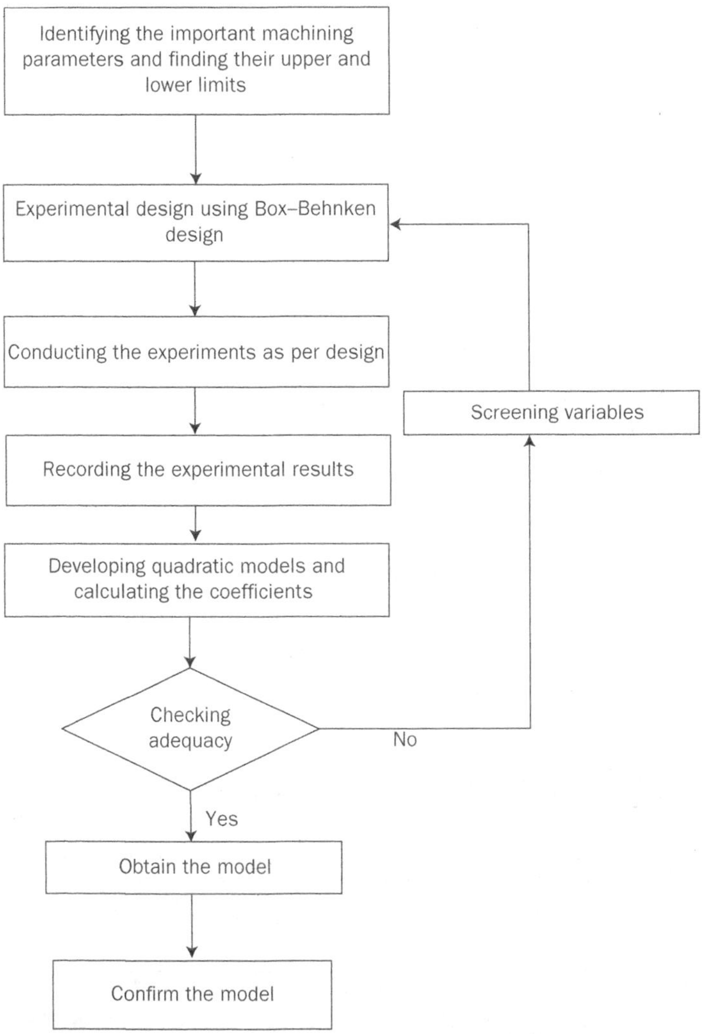

Figure 5.8 **Procedure for response surface methodology (RSM)**

5.5 Response surface model for MRR and surface roughness

The second-order response surface represents the MRR and surface roughness (R_a) can be expressed as a function of machining parameters such as pulse on-time (μs, T), electrode rotation (r.p.m., S), peak current (A, C) and flushing pressure (kg/cm^2, P). The relationship between surface roughness and machining parameters is expressed as follows:

$$R_a = \beta_0 + \beta_1 (T) + \beta_2 (S) + \beta_3 (C) + \beta_4 (P) + \beta_5 (T^2) + \beta_6 (S^2) + \beta_7 (C^2) + \beta_8 (TS) + \beta_9 (TC) + \beta_{10} (TP) + \beta_{11} (SC) + \beta_{12} (SP) + \beta_{13} (CP) \quad [5.6]$$

The model obtained for MRR is as follows:

$$\text{MRR} = 18.50945 + 0.042165{*}T + 0.093362{*}S - 2.29167{*}C - 21.08502{*}P + 2.83264\text{E-}005{*}T^2 - 2.35417\text{E-}004{*}S^2 + 0.046272{*}C^2 + 9.09223{*}P^2 - 1.82664\text{E-}004{*}T{*}S - 3.26994\text{E-}003{*}T{*}C - 0.019556{*}T{*}P + 0.014812{*}S{*}C - 0.043022{*}S{*}P + 1.45300{*}C{*}P \quad [5.7]$$

ANOVA is used to evaluate the adequacy of the model. This essentially consists of partitioning the total variation in an experiment into components ascribable to the controlled factors and error. The ANOVA result for MRR is presented in Table 5.3.

Table 5.3 Analysis of variance for material removal rate

Source	Sum of squares	d.f.	Mean square	*F*-value	Prob > *F*	Remarks
Model	4219.858	14	301.4184	18.37537	< 0.0001	significant
Residual	196.8407	12	16.40339			
Lack of Fit	160.5495	10	16.05495	0.884783	0.6390	not significant
Pure Error	36.29127	2	18.14563			
Cor Total	4416.699	26				

The Model *F*-Value of 18.38 implies that the model is significant. There is only a 0.01 per cent chance that a 'Model *F*-Value' this large could occur due to noise. The value of 'Prob > *F*' less than 0.0001 indicates that the model terms are said to be significant. The 'Lack of Fit *F*-value' of 0.88 implies the Lack of Fit is not significant relative to the pure error. The coefficient of correlation R^2 of 0.9554 indicates that the model is adequate and this model is used to navigate the design space. Similarly, the model obtained for surface roughness in EDM of WC/Co composites is as follows:

$$\text{Surface roughness} = 7.62417-4.52643\text{E-}003*T-0.017804*S+0.050665*C-0.55168*P+2.19583\text{E-}006*T^2+4.78056\text{E-}005*S^2-6.44393\text{E-}004*C^2+0.55056*P^2-2.66356\text{E-}006*T*S+1.39182\text{E-}004*T*C+8.71118\text{E-}004*T*P+4.36439\text{E-}004*S*C-5.17640\text{E-}003*S*P-0.11700*C*P$$

ANOVA results for the RSM are presented in Table 5.4.

The Model *F*-Value of 15.68 implies that the model is significant. There is only a 0.01 per cent chance that a 'Model *F*-Value' this large could occur due to noise. The value of 'Prob > *F*' less than 0.0001 indicates that the model terms are significant. The 'Lack of Fit *F*-value' of 2.06 implies the Lack of Fit is not significant relative to the pure error. The coefficient of correlation R^2 of 0.9482 indicates

Table 5.4 Result of ANOVA response surface model for surface roughness

Source	Sum of squares	d.f.	Mean square	*F*-value	Prob > *F*	Remarks
Model	8.52	14	0.61	15.68	< 0.0001	significant
Residual	0.47	12	0.039			
Lack of Fit	0.42	10	0.042	2.06	0.3713	not significant
Pure Error	0.041	2	0.021			
Cor Total	8.99	26				

that the model is adequate and this model is used to navigate the design space.

Diagnostic checking (Montgomery, 1991) of the RSM developed for MRR was performed using residual analysis and the results are shown in Figure 5.9. Residual analysis provided extra information about the model developed. The effect of the residuals which influence the MRR is analysed using a normal probability plot (Figure 5.9a). A normal probability plot is a graphical technique based on central limit theory. Lochner and Mater (1990) developed the procedure for constructing normal probability plots. In Figure 5.9(a), the data are spread approximately in a straight line, which indicates the good correlation that exists between experimental and predicted values. Figure 5.9(b) shows the residual versus predicted values, revealing minimal variation between the two. Figure 5.9(c) shows the histogram of residuals and Figure 5.9(d) shows the residuals versus

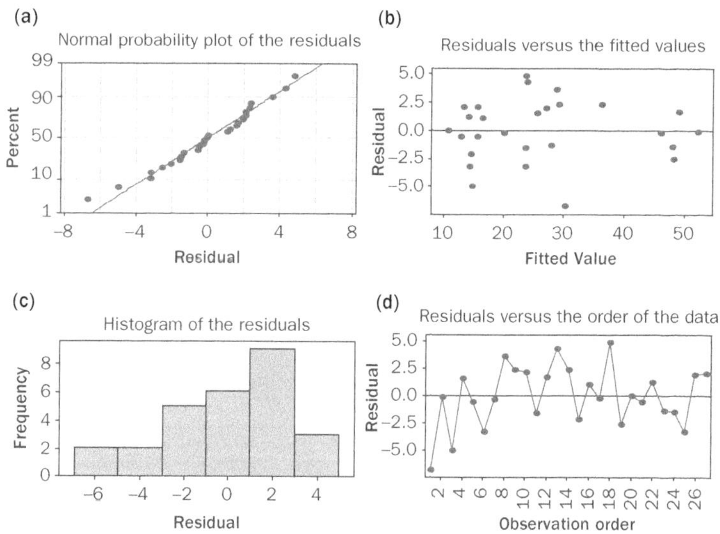

Figure 5.9 Residual graphs for material removal rate

observational order. This figure shows that a tendency to have runs of positive and negative residuals indicates a certain degree of correlation. As a whole with respect to the analysis of residuals, Figure 5.9 does not show any model inadequacy. Moreover, all the residual values fall within the control limits and hence the developed model can be used for the prediction of MRR in EDM of WC/Co composites.

Similarly, diagnostic checking of the response surface model developed for surface roughness is presented in Figure 5.10. The effect of the residuals which influence the surface roughness is analysed using a normal probability plot (Figure 5.10a); the data are spread approximately in a straight line, which indicates the good correlation that exists between the experimental and predicted values. Figure 5.10(b) shows the residual versus predicted values, revealing only minimal variation between the two. Figure 5.10(c) shows the histogram of residuals for surface roughness, and Figure 5.10(d) shows

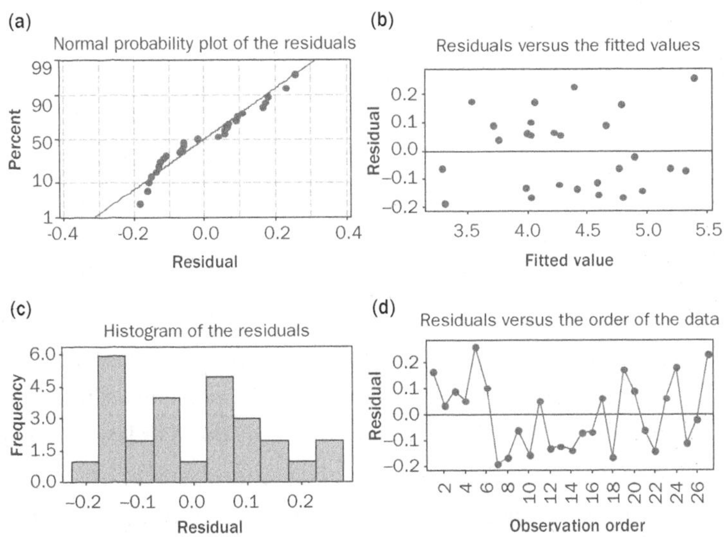

Figure 5.10 **Residual graphs for surface roughness**

the residuals versus experimental order. The tendency to have runs of positive and negative residuals indicates some level of correlation. As a whole with respect to the analysis of residuals, Figure 5.10 does not reveal any model inadequacy. Moreover, all the residual values fall within the control limits and hence the developed model is used for the prediction of surface roughness in EDM of WC/Co composites.

5.6 Results and discussion

Composites are continuously replacing more traditional materials in industry. WC/Co composites are finding applications in the manufacturing of cutting tools, dies, and other specialist tools and components owing to their hardness, strength and wear resistance over a wide range of temperatures (Tak *et al.*, 2009). In the machining of WC/Co composite materials, Liu *et al.* (2003) used CBN tools. They observed very high cutting forces and poor surface finish. For analysis of WC/Co composites and to achieve the desired shape, accurate machinability data are required. The effect of machining factors on the machinability of WC/Co composites has been studied by Jangra *et al.* (2011), who suggested that the low cobalt concentration and small grain size favors MRR. Zhang *et al.* (1997) concluded that ceramic composites are machined in EDM either by melt formation for low-melting-temperature materials or thermal spalling for refractory materials. According to Lee and Li (2001), the MRR increases with a decrease of the open circuit voltage. For low current settings, the MRR increases with increasing peak current, but then levels off when machining at higher values. In EDM, the amount of energy required to generate subsurface cracks and remove material from the surface by flake detachment is clearly less than that required to remove material more directly through melting, evaporation or dissociation, meaning that material is removed through controlled spalling (Trueman and Huddleston, 2000). Surface roughness increases with an increase in peak current, and an increase in pulse duration increases MRR. Lee and Li (2001) studied the integrity of a tungsten carbide surface after EDM. They found that surface roughness is a function of two main parameters, namely peak current and pulse duration, both of which are settings of the power supply. High peak current and/or long pulse duration produce a rough surface. At high peak current and pulse duration, numerous microcracks are observed. Removal of material from the workpiece in EDM is mainly by melting, evaporation and spalling. Spalling is the common mechanism found for some ceramic materials. This spalling effect is most often related to the generation of large microcracks (perpendicular and parallel to the top surface) generated during EDM (Panday and Jillani, 1987). The microstructure of a WC/Co-cemented carbide produced through powder metallurgy is presented in Figure 5.11.

The effect of various parameters on MRR and surface roughness has been analysed through three-dimensional

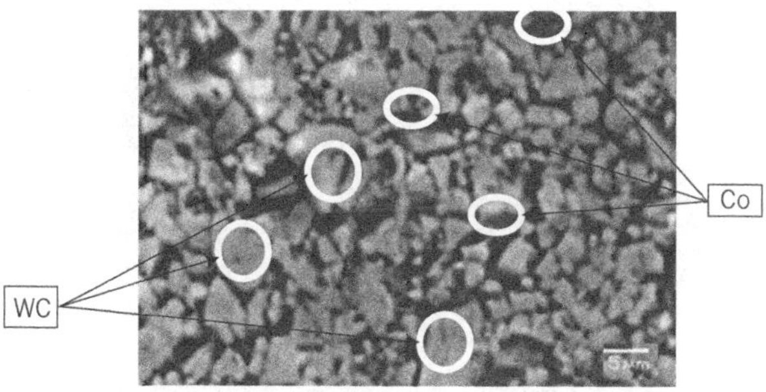

Figure 5.11 Microstructure of the as-produced composite (Kanagarajan *et al.*, 2008)

(3-D) surface graphs. The surface plot is used to visualize the response surface and is useful for establishing desirable response values and operating conditions. Such plots show how a response variable relates to two factors based on a model equation. The surface plot, which is a 3-D wireframe graph, represents the functional relationship between the response and the experimental factors. Also, the response surface helps to visualize how the response reacts to changes in experimental factors (Minitab Inc., 2003).

The effect of electrode rotation and pulse on-time on MRR in EDM of WC/Co composites is presented in Figure 5.12(a). The figure indicates that MRR decreases with an increase in the pulse on-time but increases considerably with an increase in electrode rotation speed; a higher MRR is obtained for low pulse on-time and at high electrode rotation. This is because the long pulse on-time may cause the plasma channel to expand, which reduces the energy density in EDM and is insufficient to melt and vaporize the WC/Co composite materials (Kuppan and Narayanan, 2006). The effect of peak current and pulse on-time on

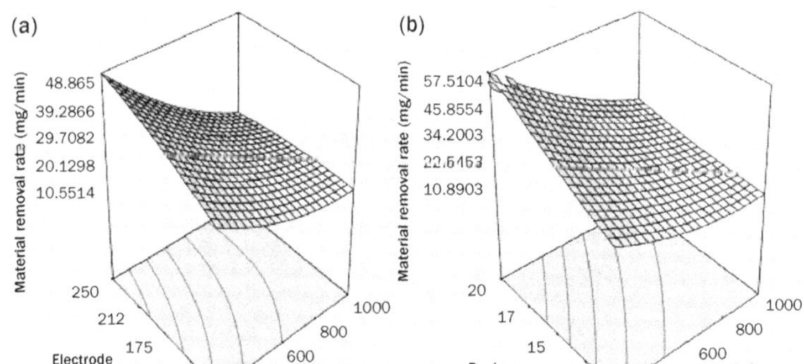

Figure 5.12 (a) Effect of electrode rotation and pulse on-time on MRR, (b) effect of peak current and pulse on-time on MRR

MRR is presented in Figure 5.12(b), which indicates that an increase of peak current increases MRR, whereas pulse on-time shows the reverse trend.

Figure 5.13(a) shows a 3-D graph of flushing pressure and pulse on-time on MRR. An increase of flushing pressure increases MRR, in contrast to the effect of pulse on-time. The increased MRR with increasing flushing pressure is due to the particles in the machining gap being evacuated more efficiently (Ghoreishi and Atkinson, 2002). The effect of peak current and electrode rotation on machining of WC/Co composites with respect to MRR is presented in Figure 5.13(b), showing that an increase of peak current and electrode rotation increases MRR. The increase of peak current increases the energy released, such that material removal is easier and deeper. High electrode rotation increases MRR: more material is removed in less time.

The interaction effect between flushing pressure and electrode rotation on MRR is presented in Figure 5.14(a) as a 3-D response graph. The figure indicates that increasing flushing pressure and electrode rotation increase MRR in

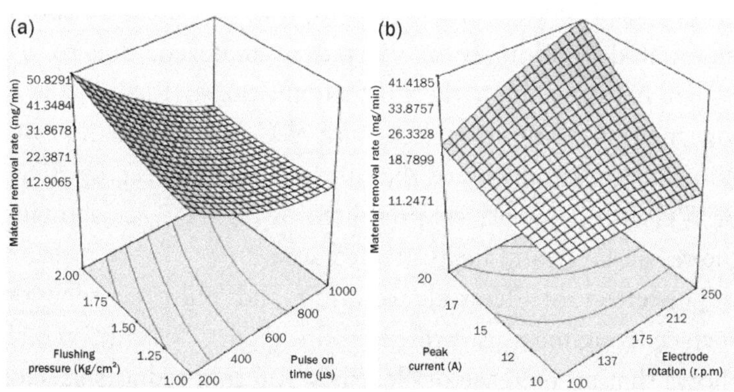

Figure 5.13 **(a) Effect of flushing pressure and pulse on-time on MRR, (b) effect of peak current and electrode rotation on MRR**

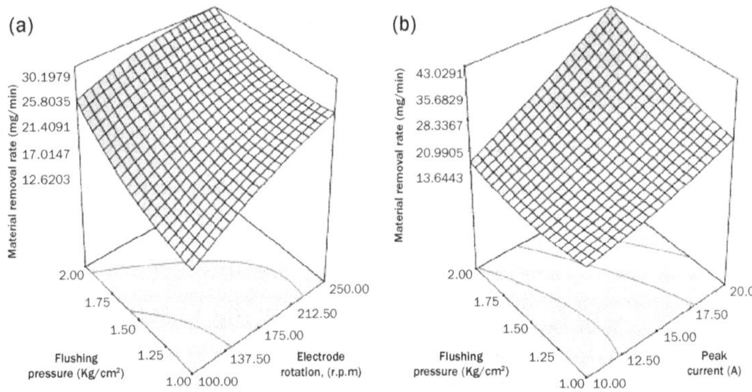

Figure 5.14 (a) Effect of flushing pressure and electrode rotation on MRR, (b) effect of flushing pressure and peak current on MRR

machining of WC/Co composites as discussed earlier. The effect of flushing pressure (*P*) and peak current (*C*) on MRR is presented in Figure 5.14(b) as a 3-D response graph. The figure indicates that the maximum MRR is obtained at 20 A and a flushing pressure of 2.0 kg/cm^2. With increasing current intensity, the energy required for EDM increases so that machining is easier and the discharge craters become deeper and wider, thus contributing to greater cobalt material debonding from the parent material, due to the cobalt having a high thermal expansion coefficient and a low melting point compared with WC (Kanagarajan *et al.*, 2008b). A micrograph of the machined workpiece by EDM of WC/Co composites is presented in Figure 5.15, showing the recast layer and heat-affected zone.

The effect of electrode rotation and pulse on-time on surface roughness is presented in Figure 5.16(a), which shows that an increase of electrode rotation reduces surface roughness. The reason for this is that an increase of electrode rotation eases the machining, which in turn reduces surface roughness. The increase of pulse on-time increases the spark,

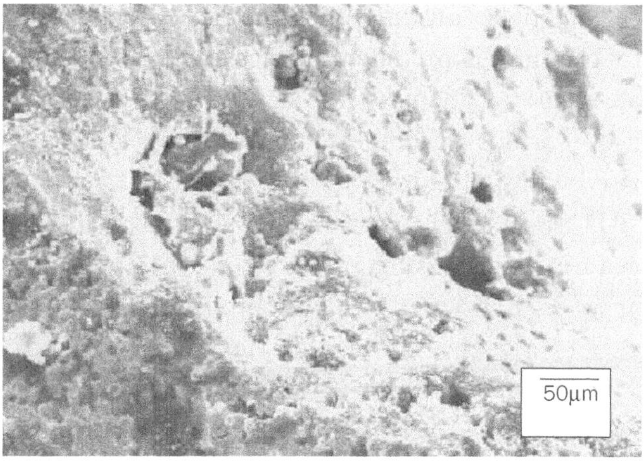

Figure 5.15 Micrograph of the machined workpiece with heat-affected zone and recast layer

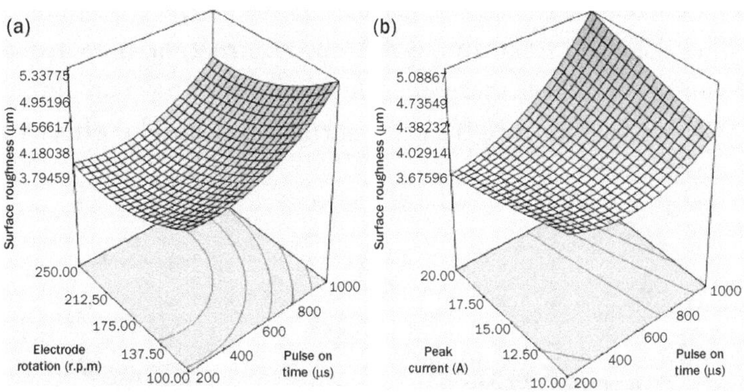

Figure 5.16 (a) Effect of electrode rotation and pulse on-time on surface roughness, (b) effect of peak current and pulse on-time on surface roughness

which results in an increased MRR. Consequently, the craters produced on the machined surface are broader and deeper. This affects the surface finish: with shorter duration of sparks, the surface finish is better. The effect of peak current and pulse on-time on the surface roughness in an EDM workpiece is presented in Figure 5.16(b). The results indicate that an increase of current and pulse on-time increases surface roughness, but not as noticeably as compared with pulse on-time.

A 3-D surface graph showing the interaction effect between flushing pressure and electrode rotation on surface roughness in EDM of WC/Co composites is presented in Figure 5.17(a). The figure indicates that an increase of electrode rotation reduces the surface roughness in machining of WC/Co composites. And an increase in flushing pressure improves the surface finish. With increasing electrode rotation, the ignition time delay increases, reducing the energy transferred through the individual discharges for material removal. This diminishes the dimensions of craters, giving a better surface roughness (Ghoreishi and Atkinson, 2002). By contrast, an increase in flushing pressure increases flushing, making machining easier and producing a better surface finish. The effect of

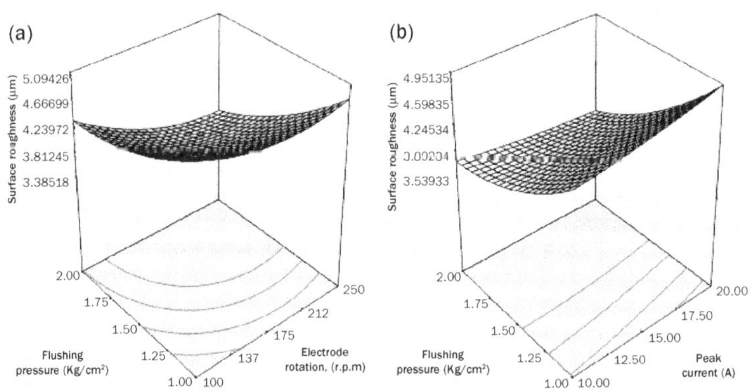

Figure 5.17 (a) Effect of flushing pressure and electrode rotation on surface roughness, (b) effect of flushing pressure and peak current on surface roughness

flushing pressure and peak current on the surface roughness in EDM is presented in Figure 5.17(b). This indicates that an increase of peak current increases surface roughness, whereas an increase of flushing pressure reduces surface roughness.

The effect of flushing pressure and pulse on-time on surface roughness is shown in a 3-D surface graph in Figure 5.18(a). As already discussed, the increase of flushing pressure and electrode rotation reduce surface roughness. The effect of peak current and flushing pressure on surface roughness is presented in Figure 5.18(b), showing that an increase of flushing pressure reduces surface roughness, whereas an increase in peak current increases surface roughness. The increase in current produces an increase in surface roughness due to increased discharge energy, which subsequently causes larger craters on the surface of the body. The machined surface obtained on WC/Co composite is presented in Figure 5.19.

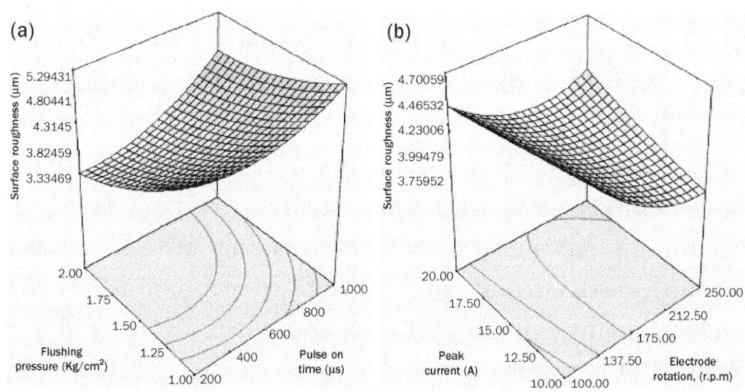

Figure 5.18 (a) Effect of flushing pressure and pulse on-time on surface roughness, (b) effect of peak current and electrode rotation on surface roughness

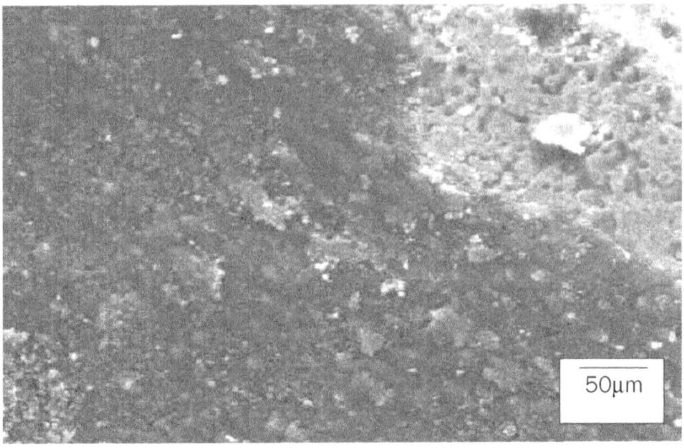

Figure 5.19 Surface produced on machining of WC/Co composites

5.7 Summary

In this chapter we have investigated EDM of WC/Co-cemented carbide, an important composite material used in the manufacture of cutting tools, dies and other specialist tools, first presenting a general overview of the EDM process. The effects of various machining parameters which influence the EDM process in the machining of WC/Co composites have been analysed using response surface analysis. The experiments were carried out using a Box–Behnken experimental strategy. The effect of machining parameters such as flushing pressure, pulse on-time, peak current and electrode rotation on MRR and surface quality of the WC/Co composites was analysed using 3-D response graphs. Further detailed research is underway to analyse other parameters which influence EDM.

Acknowledgements

We thank Drs D. Kanakarajan and R. Karthikeyan, Annamalai University, Chidambaram, for their help in the preparation of this chapter.

References

Anees A K, Mansoor A K (1996), 'Box-Behnken design for the optimization of formulation variables of indomethacin coprecipitates with polymer mixtures,' *International Journal of Pharmaceutics* **131**: 9–17.

Benardos P G, Vosniakos G C (2003), 'Predicting surface roughness in machining: a review', *International Journal of Machine Tools and Manufacture*, **43**: 833–44.

Box G E P, Behnken D W (1960), 'Some new three level designs for the study of quantitative variables', *Technometrics*, **2**: 455–75.

Chen S I, Hsu Q C (2003), 'Studies on electrical-discharge machining of non-contact seal face grooves', *Journal of Materials Processing Technology* **140**: 363–7.

Chen Y, Mahdavian S M (1999), 'Parametric study into erosion wear in a computer numerical controlled electro-discharge machining process', *Journal of Wear* **236**: 350–4.

Francis E H T, Sumit K S, Mannan M A (2002), 'Topography of the flank wear surface', *Journal of Materials Processing Technology*, **120**: 243–8.

George K, Ragunath P M, Manocha P K, Ashish L M, Warrier M (2004), 'EDM machining of carboncarbon composite – a Taguchi approach', *Journal of Materials Processing Technology*, **145**(3): 66–71.

Ghoreishi M, Atkinson J (2002), 'Comparative experimental study of machining characteristics in vibratory, rotory and vibro rotory electro discharge machining', *Journal of Materials Processing Technology*, **120**: 374–84.

Godfrey C, Onwubolu S K (2006), 'Response surface methodology-based approach to CNC drilling operations', *Journal of Materials Processing Technology*, **171**: 41–7.

Jangra K, Grover S, Chan F T S, Aggarwal A (2011), 'Digraph and matrix method to evaluate the machinability of tungsten carbide composite with wire EDM', *International Journal of Advance Manufacturing Technology*, **56**: 959–74.

Janmanee P, Muttamara A (2010), 'Performance of difference electrode materials in electrical discharge machining of tungsten carbide', *Energy Research Journal* **1**: 87–90.

Kanagarajan D, Karthikeyan R, Palanikumar K, Paulo Davim J (2008a), 'Optimization of electrical discharge machining characteristics of WC/Co composites using non-dominated sorting genetic algorithm (NSGA-II)', *International Journal of Advanced Manufacturing Technology*, **36**: 1124–32.

Kanagarajan D, Karthikeyan R, Palanikumar K, Davim P J (2009), 'Application of goal programming technique for Electro Discharge Machining (EDM) characteristics of cemented carbide (WC/Co)', *International Journal of Materials and Product Technology*, **35**: 216–27.

Kanagarajan D, Karthikeyan R, Palanikumar K, Sivaraj P (2008b), 'Influence of process parameters on electric discharge machining of WC/30%Co composites', *Proceedings of the Institution of Mechanical Engineers, Part B: Journal of Engineering Manufacture*, **222**: 807–15.

Kuppan R A, Narayanan S (2006), 'Experimental investigation into electrical discharge deep hole drilling of INCONAL 718 using graphite electrode', *22nd AIMTDR Conference IIT Roorke, India.*

Kwak J-s (2005), 'Application of Taguchi and response surface methodologies for geometric error in surface grinding process', *International Journal of Machine Tools and Manufacture*, **45**: 327–34.

Lajis M A, Radzi H C D M, Amin A K M N (2009), 'The implementation of Taguchi method on EDM process of tungsten carbide', *European Journal of Scientific Research* **26**: 609–17.

Lauwers B, Vleugels J, Malek O, Brans K, Liu K (2012), 'Electrical discharge machining of composites', In: *Machining Technology for Composite Materials*, Woodhead Publishing, in press.

Lee S H, Li X P (2001), 'Study of the effect of machining parameters on the machining characteristics in electrical discharge machining of tungsten carbide', *Journal of Material Process Technology* **115**: 344–50.

Lin J L, Lin C L (2002), 'The use of orthogonal array with grey relational analysis to optimize the electrical discharge machining process with multiple performance characteristics', *International Journal of Machining Tools Manufacture*, **42**: 237–44.

Liu K, Li X P, Rahman M (2003), 'Characteristics of high speed micro cutting of tungsten carbide', *Journal of Material Processing Technology*, **140**: 352–7.

Llanes L, Casas B, Idanez E, Marsal M, Anglada M (2004), 'Surface integrity effects on the fracture resistance of electrical-discharge-machined WC–Co cemented carbides', *Journal of American Ceramic Society*, **87**: 1687–93.

Lochner R H, Mater J E (1990) *Designing for Quality*. London: Chapman & Hall.

Luis C J, Puertas I, Villa G (2005), 'Material removal rate and electrode wear study on the EDM of silicon carbide', *Journal of Material Processing Technology*, **164(165)**: 889–96.

Montgomery D C (1991) *Design and Analysis of Experiments*. New York: John Wiley and Sons.

Mahdavinejad R A, Mahdavinejad A (2005), 'ED machining of WC–Co', *Journal of Materials Processing Technology*, **162–163**: 637–43.

Minitab Inc (2003) *Minitab® Statistical Software*.

Palanikumar K (2008), 'Application of Taguchi and response surface methodologies for surface roughness in machining glass fiber reinforced plastics by PCD tooling', *International Journal of Advanced Manufacturing Technology*, **36**: 19–27.

Panday P C, Jillani S T (1987), 'Electrical machining characteristics of cemented carbides', *Wear*, **116**: 77–88.

Puertas I, Luis C J (2004), 'A study on the electrical discharge machining of conductive ceramics', *Journal of Materials Processing Technology*, **153**: 1033–8.

Puertas I, Luis C J, Álvarez L (2004), 'Analysis of the influence of EDM parameters onsurface quality, MRR and EW of

WC–Co', *Journal of Materials Processing Technology*, **153–154**: 1026–32.

Hecker R L Liang S Y (2003), 'Predictive modeling of surface roughness in grinding', *International Journal of Machine Tools and Manufacture*, **43**: 755–61.

Sahin Y, Motorcu A R (2008), 'Surface roughness model in machining hardened steel with cubic boron nitride cutting tool', *International Journal of Refractory Metals and Hard Materials*, **26**: 84–90.

Sano S, Suzuki K, Pan W, Iwai M, Murakami Y, Uematsu T (2007), 'Forming fine V-grooves on a tungsten carbide workpiece with a PCD electrode by EDM', *Key Engineering Materials*, **329**: 631–6.

Tak H-S, Ha C-S, Kim D-H, Lee H-J, Lee H-J, Kang M-C (2009), 'Comparative study on discharge conditions in micro-hole electrical discharge machining of tungsten carbide (WC-Co) material', *Transactions of Nonferrous Met. Soc. China* **19**: s114–s118.

Tomadi S H, Hassan M A, Hamedon Z, Daud R, Khalid A G (2009), 'Analysis of the influence of EDM parameters on surface quality, material removal rate and electrode wear of tungsten carbide', In: *Proceedings of the International Multi-Conference of Engineers and Computer Scientists (IMECS)*, vol. II, Hong Kong.

Trueman C S, Huddleston J (2000), 'Material removal by spalling during EDM of ceramics', *Journal of the European Ceramics Society*, **20**: 1629–35.

Zhang J, Lee T C, Lau W S (1997), 'Study on the electro-discharge machining of a hot pressed aluminum oxide based ceramic', *Journal of Materials Processing Technology*, **63**: 908–12.

6

Conventional and unconventional hole making in metal matrix composites

Inderdeep Singh[1], *Sarbjit Singh*[1] *and Abhishek Singh*[1]

Abstract: Hole making in advanced materials such as metal matrix composites (MMCs) is an essential prerequisite to facilitate their assembly. The secondary processing of MMCs is thus unavoidable in the overall product development cycle. Here we address the issues and challenges in conventional and unconventional hole-making techniques in MMCs. There are a plethora of challenges associated with achieving good-quality cost-effective hole making in MMCs. The conventional (drilling) and unconventional (ultrasonic and electrical discharge machining) techniques of hole making in MMCs are discussed, and the results of previous research are outlined. The chapter provides insight into the fundamental aspects of hole making in MMCs.

Key words: Drilling, electric discharge machining, hole making, MMCs, ultrasonic machining.

[1] Department of Mechanical and Industrial Engineering, Indian Institute of Technology, Roorkee – 247667, India.

Published by Woodhead Publishing Limited, 2013

6.1 Introduction

Metal matrix composites (MMCs) have been the subject of research interest since the 1970s. MMCs possess attributes that make them advantageous over non-composite materials, including a combination of enhanced properties for structural applications, aerospace, automobile and household products, for example increased strength to weight ratio, high toughness and impact strength, high elastic modulus, high surface durability, low sensitivity to thermal shocks, low sensitivity to surface flaws and good wear resistance. Such improved characteristics make MMCs ideal for advanced engineering applications. 'Design for Machining' is an important criterion when ascertaining whether these materials can be appropriately used, as machining represents the leading cost associated with the manufacturing of a product. Although efforts have been made to develop near net shape manufacturing of components, some secondary machining processes such as drilling need to be performed for joining and assembly purposes.

6.2 Development of the secondary processes for MMCs: 'need of the hour'

MMCs were developed in the early 1970s and the primary support for their development was from the aerospace industry. At present, more than 80 per cent of MMCs are used for ground transportation, so primary support for the development of MMCs is now from the transportation industry. The transportation industry is looking for efficient and cost-effective methods for the development and assembly of MMCs. Therefore, efforts have been made to develop near net shape manufacturing techniques for MMC products. However, in many structural and engineering applications, machining is essential to complete the assembly process. Therefore, for joining and assembly, machining processes such as drilling are often unavoidable. Drilling is often the last manufacturing process to be performed on a part before assembly. Hole making in MMCs by unconventional methods such as electrical discharge machining (EDM) necessitates higher set-up costs and results in lower productivity (lower material removal rate, MRR). Therefore, conventional drilling is the most widely acceptable and frequently practised machining operation for hole making in advanced materials such as MMCs.

6.3 Classification of MMCs based on the matrix material

MMCs are classified into different categories depending upon the matrix material, such as aluminium matrix composites, magnesium matrix composites, titanium matrix composites, copper matrix composites and super alloy matrix composites (Figure 6.1). Aluminium matrix composites are of major research interest as they are widely used in the automobile and transportation industries. Their use in industry is due to the fact that reinforcement compounds such as SiC, Al_2O_3 and B_4C can be mixed easily in molten aluminium rapidly and cost-effectively. Magnesium matrix composites also have similar advantages, but due to limitations in fabrication and lower thermal conductivity, they are not widely used. Magnesium matrix composites have been developed for the space industry to take advantage of the low weight of magnesium as compared with aluminium. Titanium alloys are used as matrix material in fabricating MMCs due to their good strength at elevated temperatures and excellent corrosion resistance. Compared with aluminium, titanium alloys retain their strength at higher

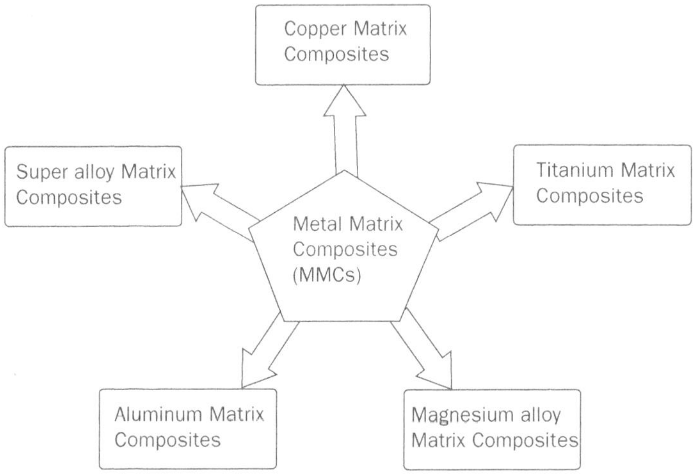

Figure 6.1 Classification of MMCs based on matrix material

temperature, which is advantageous in manufacturing aircraft and missile structures, where operating speeds are very high. However, the main problem lies with processing of highly reactive titanium with reinforcement materials. Fibre-based titanium MMCs are widely used in developing aircraft structures. Copper matrix composites are superior in terms of thermal conductivity and high-temperature strength properties compared with other MMCs. Super alloys are commonly used as candidate materials for manufacturing gas turbine blades, where they operate at higher temperature and speeds.

6.4 Fabrication processes for MMCs

The manufacturing of MMCs is generally classified into two categories: primary processing and secondary processing. The preliminary production of MMCs is via primary processing, in which the various constituents of the composites are combined but this may or may not lead to the final product or concluding microstructure. The final shape and microstructure of the composites are altered by secondary processing, such as heat treatment, machining and drilling. The selection of either primary or secondary processing depends upon the type of matrix, type of reinforcement, percentage of reinforcement, shape and size of reinforcement, and its mechanical and chemical characteristics. The primary processing of MMCs (Figure 6.2) is generally classified into two categories based upon the phase of the matrix (the matrix is either in solid phase or in liquid phase during addition of the reinforcement).

Liquid phase processing methods are more efficient and economic than solid phase methods. The majority of MMCs are fabricated via a liquid metallurgy route due to cost advantages. Stir-casting is a popular technique because of economic considerations and is the simplest method of processing MMCs. Generally, stir casting involves the addition of reinforcement material into the metal matrix and the mixture is stirred at a set speed for a specific period. The mixture is stirred to form a vortex at the surface of the melt. Stirring continues for a few minutes before the slurry is cast. The vortex method is used to maintain the even distribution of the reinforcement particles in the matrix alloy. Processing

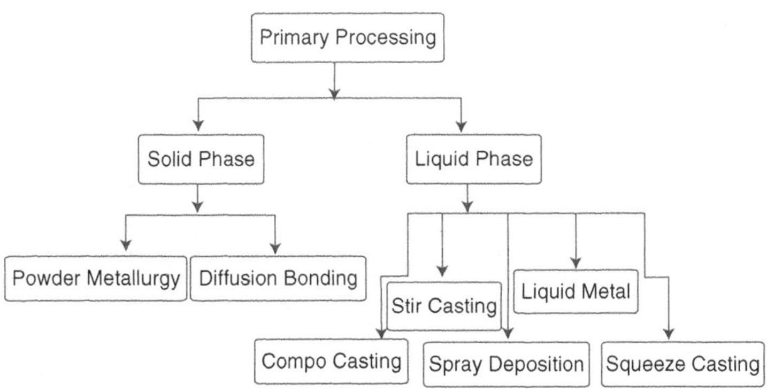

Figure 6.2 Classification of primary processing of MMCs

variables such as holding temperature, stirring speed, and size and position of the impeller in the melt are important factors to be considered in producing good-quality MMC castings.

6.5 Conventional and unconventional machining processes

Comparative data relating to the various conventional and unconventional processes of hole making are presented in Table 6.1. These data are helpful in selecting a hole-making process with respect to the application of the product. One specific hole-making process that is appropriate under the

Table 6.1 General comparative study of the different hole-making processes for various materials

Comparison	Process				
	Electro-discharge machining	**Electro-discharge drilling**	**Electro-chemical machining**	**Ultrasonic machining**	**Conventional drilling**
Experimental time	Long	Medium	Medium	Long	Short
Material	Conductive only	Conductive only	Conductive only	Conductive and non-conductive	Conductive and non-conductive
Burr formation	No	No	No	No	Yes
Material removal rate	Slow	Fast	Fast	Slow	Very fast
Tool wear rate	Maximum	Minimum	Minimum	Extensive	Maximum
Accuracy level	Normal	High	Low	High	Very high
Surface roughness	Poor	Average	Good	Average	Good
Cutting force and torque	No	No	No	Yes with vibration	Yes
Thermal stresses	Negligible	Negligible	No	Negligible	Extreme
Residual stresses	Yes	Negligible	No	Negligible	Yes
Tool cost	Minimum	Minimum	Minimum	Minimum	Maximum
Power consumption	Medium	Medium	High	Low	Low

Published by Woodhead Publishing Limited, 2013

given conditions for one material may not be equally effective for another. Therefore, a database of process selection for a given material depending upon product quality and service requirements is essential.

6.6 Conventional secondary processing of MMCs

Secondary processing includes a process or combination of processes performed on the MMCs for making a useful product. Secondary processing can be broadly classified into two categories (Figure 6.3):

- processes to make the shape of the product
- processes to alter the properties of the product.

The first category includes material removal (machining, shaping, drilling, milling, etc.) or material deformation

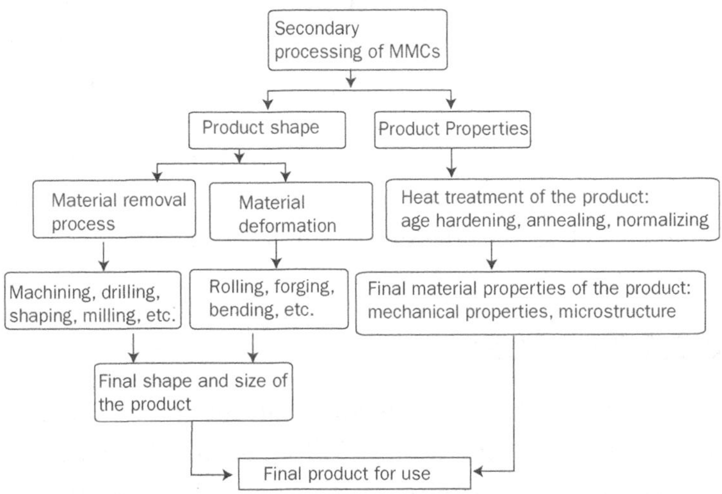

Figure 6.3 Classification of secondary processing of MMCs

Published by Woodhead Publishing Limited, 2013

(forging, bending, rolling, etc.) processes. In the former (conventional machining process), the final shape of the product is made by removing extra material with the help of a cutting tool with relative motion between the cutting tool and the workpiece. Plastic deformation of the material takes place during the material deformation processes to alter the properties of the product. Plastic deformation may be carried out at elevated temperature (above recrystallization temperature) or at room temperature under the application of an applied load. The internal stresses induced by the plastic deformation processes are greater than those induced by material removal processes.

6.7 Hole making in MMCs

Hole making in MMCs with conventional tooling led to a novel set of challenges for researchers. The abrasive nature of the reinforcement causes high tool wear rate, increases the cutting forces and decreases the surface quality of the drilled hole. The various machine tool parameters (feed rate, cutting speed) and cutting tool parameters (tool material, point angle) have been investigated to obtain an optimum combination of parameters for optimized output quality characteristics. The detailed behaviour of input parameters and output responses is discussed briefly in the subsequent sections.

6.7.1 Tool wear

Tool wear is the major factor that influences the hole quality characteristics during drilling of MMCs, where the quality of the machined part produced is of primary importance. The abrasive nature of the ceramic reinforcement, such as SiC and Al_2O_3, means that it acts as a grinding tool and abrades the cutting edge of the cutting tool. Tool wear is influenced by the machine tool parameters and by the material properties (either workpiece or cutting tool). The machine tool parameters include feed rate, cutting speed, vibration-assisted drilling, type of cooling, lubrication, etc. Certain uncontrollable parameters such as environmental conditions, temperature, humidity, vibration induced in the machine structure, and mental or physical fatigue of the worker also play a significant role in tool wear. The material of the cutting tool, geometry of the cutting tool, type of matrix, type of reinforcement, shape and size of reinforcement, and percentage of reinforcement are important parameters related to material properties, and have a significant role in characterizing tool wear. Cutting tool performance while drilling of MMCs is generally similar to that of conventional tool materials such as high-speed steel (HSS). A large variety of tool materials (Figure 6.4) are investigated here, such as HSS, high cobalt HSS, titanium nitride, tungsten carbide (WC), carbide-tipped HSS, solid carbide and polycrystalline diamond (PCD). Among these cutting tool materials, uncoated HSS drills are

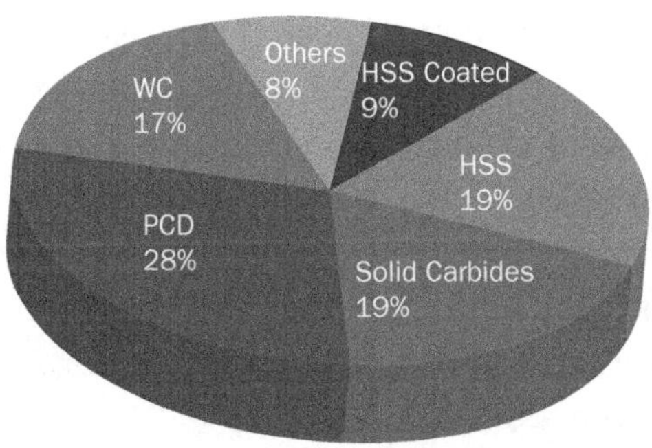

Figure 6.4 **Tool materials used for drilling of MMCs**

not suitable for drilling in MMCs because of rapid tool wear (above that recommended by the ISO) and poor surface finish. High cobalt HSS and titanium nitride show improved characteristics, but the best results are obtained using solid carbide and PCD drills (Monaghan and Reilly, 1992). The superior performance of solid carbide and PCD drills is because of high abrasion resistance and high thermal conductivity, which results in a lower cutting temperature (Mubaraki *et al.*, 1995). In contrast to monolithic materials, in the case of drilling of MMCs tool wear is inversely proportional to feed rate (Morin *et al.*, 1995). This is because the probability of interactions between the cutting tool and ceramic reinforcement decreases with an increase in feed rate. Therefore, fewer abrasive particles abrade the cutting tool material, causing a decrease in tool wear. No uniform pattern of tool wear is observed with a change in cutting speed. Flank wear and crater wear are observed on HSS, WC and coated HSS but only crater wear is observed on solid carbide drills and PCD drills. The predominant mechanism of wear is abrasive wear, which leads to an increase in feed force (thrust force) with an increase in tool wear. Built-up edges (BUEs) are observed on solid carbide drills and PCD drills on the chisel edge and flank edges. BUEs have both advantages and disadvantages. A BUE protects the cutting edge from wear as it makes an extra cutting edge on the main cutting edge. The developed BUE also causes an increase in feed force and decrease in the surface finish of the workpiece. It can be concluded from the parametric evaluation of the machine tool parameters that cutting time and feed rate are the predominant factors that effect tool wear during drilling of MMCs (Davim, 2003).

6.7.2 Cutting forces

The hard ceramic reinforcing component in MMCs makes them difficult to machine and an attempt to do so leads to high drilling thrust force and torque. The thrust force and torque, and therefore the power required are related to the magnitude of the drill tool wear. The thrust force for HSS drills increases with increasing feed rate at a particular cutting speed. The lowest forces are recorded for the test performed using PCD drills, followed by carbide and HSS tools. The order of these results indicates clearly that tool hardness has a major influence on the efficiency of the drilling process when machining MMCs (Monaghan and Reilly, 1992). Torque shows similar trends. Cutting speed has no significant effect on tool wear or on drilling forces. A linear relationship between both thrust and torque versus flank wear has been observed, so thrust or torque may be measured to give an indication of wear of the drills (Morin *et al.*, 1995). The linear relationship between torque and tool wear of the drill implies a linear variation of specific cutting energy with flank wear. PCD drills produce lower drilling forces than those generated by carbide-tipped drills. Drilling forces also depend on the hardness of the reinforcement. As the volume fraction of the reinforcement increases significantly, regardless of the tool material and workpiece material, both thrust force and torque are highly dependent on feed rate (Ramulu *et al.*, 2002). Cutting speed was found to have an insignificant influence on the magnitude of drilling forces. A significant reduction in cutting forces (Basavarajappa *et al.*, 2008) is observed with the addition of 1–3 per cent graphite (hybrid composites). This is attributed to the solid lubricating property of the graphite particles. The graphite particles reduce the interfacial friction between the tool and workpiece and lower the shear flow stress. The addition of graphite decreases the hardness and strength of the hybrid composites, thereby improving machinability, except in ductile material, which may generate a BUE, burr and poor surface finish. The most significant process parameters that significantly affect the cutting forces are feed rate and tool material. Cutting speed is insignificant in predicting the cutting forces during drilling of MMCs.

6.7.3 Surface roughness

The surface roughness produced in drilling of MMCs comprises three parts: one originates inherently as feed marks depending upon tool geometry and the magnitude of the cutting speed and feed rate, the second one appears gradually due to deterioration of the cutting edges and vibration, and the third is due to particle pull out and damage (Riaz Ahamed *et al.*, 2009). Therefore, machine tool parameters and material properties such as cutting speed, feed rate, geometry of the cutting tool, type of material, type of chip produced, work material and vibration of the machining system all have significant effects on the surface roughness of the drilled hole wall.

Surface roughness increases with an increase in feed rate but decreases with a decrease in cutting edge radius, which may occur due to tool wear. At constant cutting speed, the surface roughness of the holes increases with increasing feed rate but does not change significantly with cutting speed. However, some researchers have suggested that surface roughness increases with an increase in feed rate, and decreases when cutting speed is increased. The lowest surface roughness is reported at the lowest feed rate and highest cutting speed. The decrease in surface roughness with increase in cutting speed can be attributed to the increased burnishing and honing effect produced by the trapped abrasive particles between the tool and the drilled hole wall (Basavarajappa *et al.*, 2007). The increase in surface roughness with increase in feed rate results because the contact time between the tool and workpiece decreases, thus reducing the burnishing effect. The increase in feed rate also causes more serious BUEs on the tip of the cutting tool, which lead to an increase in surface roughness of the drilled hole wall. A high feed rate increases the heat generated and hence temperature of the cutting, which accelerates diffusion and adhesion wear in the thermally weakened tool. Tool chatter also increases with an increase in feed rate, which leads to increased surface roughness. Taguchi analysis of the effect of process variables on the surface roughness of MMCs shows that feed rate is an important parameter which has a significant influence on surface roughness (43 per cent) followed by cutting speed (41 per cent). The interaction between cutting speed and feed rate (14 per cent) is the most important interaction in hole surface analysis (Davim, 2003).

Cutting tool hardness is an important parameter in predicting surface roughness. Hard-carbide tools produce a better surface finish compared with that achieved when using HSS and TiN-coated HSS drills (Monaghan and Reilly, 1992). The hardness of the matrix exerts a significant influence on the surface finish texture of the MMC. When drilled with hard drills, such as solid carbide drills, poor surface roughness is associated with drilling the material under softer conditions. By contrast, HSS drills produce a better surface finish in the softer matrix. In each case, there is similarity between the result for point angles and surface roughness. With increasing point angle of all the drills, there is a decrease in surface roughness (Kilickap, 2010). The inclusion of graphite within hybrid composites leads to poor surface finish; graphite, being less dense, acts as a weak spot. When the tool passes over these regions, the crushed graphite particles form a deep valley and hence increase the surface roughness of the material. The removed graphite particles become smeared on the machined surface. This smeared graphite layer also lowers the coefficient of friction. In addition, the burnishing effect of the trapped ceramic abrasives is reduced when graphite is encountered at the surface of the workpiece material (Basavarajappa *et al.*, 2007).

6.8 Unconventional machining of MMCs

There are clearly many issues and challenges with conventional hole making in MMCs and these need to be addressed if we are to increase the range of application of various types of MMCs. In this regard, with innovations in the field of advanced materials, various unconventional machining processes are being developed. The processing of MMCs via unconventional processes is one of the major innovations in the field of manufacturing science and engineering. To justify the application spectrum of MMCs, it is imperative to characterize their manufacturability, in addition to their usual mechanical, physical and chemical properties. Advanced materials (such as MMCs) present researchers and engineers with a plethora of challenges in ascertaining their high-quality cost-effective processing. To overcome these challenges, a number of innovative material removal processes have been developed to meet the standards of commercial utilization (Ghosh and Mallik, 2001). These new methods are known as unconventional machining processes.

Unconventional machining processes are broadly classified as mechanical, chemical or thermal, depending upon the mechanism of material removal (Table 6.2). For hole making in MMCs, ultrasonic machining (USM) (mechanical), electro-chemical machining (chemical), and electro-discharge

Table 6.2 Classification of unconventional machining processes based on the energy scheme used for material removal

Energy type	Processes	Medium	Basic mechanism	Hole length	
				(<15–20 mm)	(>15–20 mm)
Mechanical processes	Ultrasonic machining (USM)	Abrasive/water	Abrasion	Good	Poor
	Water jet machining (WJM)	Water	Erosion	Fair	Poor
	Abrasive jet machining (AJM)	Abrasive/gas	Erosion	Fair	Poor
Chemical processes	Electrochemical machining (ECM)	Electrolyte	Ion displacement	Good	Good
	Electrochemical grinding (ECG) etc.	Electrolyte	Plastic shear, ion displacement	Not possible	Not possible
Thermal processes	Electro-discharge machining (EDM)	Dielectric	Fusion and vaporization	Good	Good
	Electro-discharge drilling (EDD)	Dielectric	Fusion and vaporization	Good	Good
	Plasma beam machining (PBM)	Gas	Ionized substance, plasma	Fair	Not possible
	Laser beam machining (LBM)	Air	Powerful radiation, photons	Fair	Poor
	Electron beam machining (EBM) etc.	Vacuum	High-speed electrons, electron beam	Not possible	Poor

machining (EDM) or electro-discharge drilling (EDD) (thermal) processes have been developed and described.

Figure 6.5 shows the latest figures regarding the application of unconventional machining to the processing of conventional and advanced materials. More than 200 research publications have appeared on unconventional machining of materials in the last decade. The major emphasis of this research work has been on the steel industry, used for making dies, cutting tools, profile cutting, etc. With an increase in the use of advanced materials such as MMCs in the transportation sector, aerospace components, sports and household products, research efforts in the field of machining of MMCs have undergone unprecedented acceleration in recent years.

Figure 6.6 shows the distribution of research studies conducted in the field of unconventional machining of MMCs. The largest proportion concern thermal processes (EDM, EDD) for machining of MMCs (approximately 50 per cent), followed by chemical and mechanical processes (31 and 21

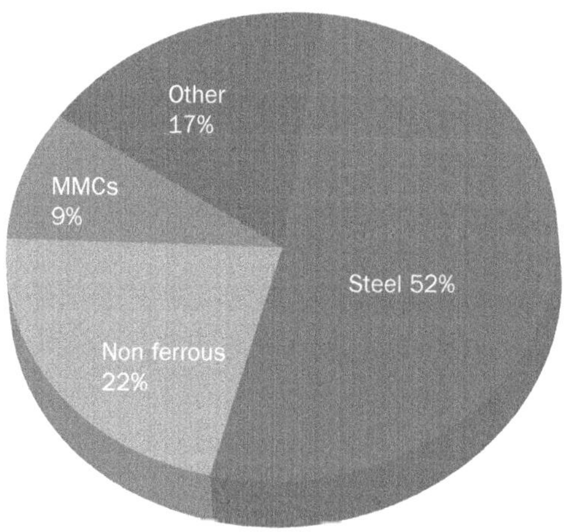

Figure 6.5 Research studies conducted on unconventional machining of various materials

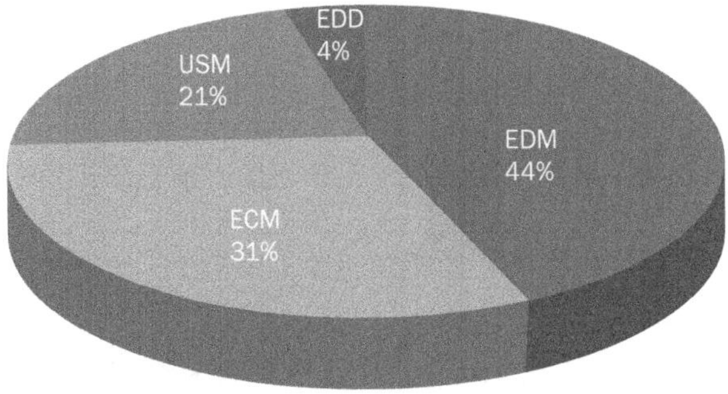

Figure 6.6 Research studies conducted on various unconventional hole-making processes

per cent, respectively). Unconventional machining processes have superior performance over conventional machining operations because of the high degree of automation possible, superior material properties obtained and ability to generate complex profiles with precision.

6.8.1 Ultrasonic machining

USM was introduced in the early 1940s. It is a mechanical material removal process that is mainly used for impact erosion of electrically conductive, non-conductive quartz to obtain straight holes, blind holes, round holes and other complex shapes. The basic principle of USM is based on material removal from the workpiece by repetitive impact of abrasive particles carried by the fluid in the form-slurry under the action of a mirror-shaped vibrating tool. The development of advanced materials such as MMCs has increased the application spectrum of USM. Figure 6.7 details the USM process. A vibrating tool oscillates at low amplitude (25–100 μm) and high frequency (15–30 kHz) (Pandey and Shan, 1980). The abrasive slurry (with abrasives

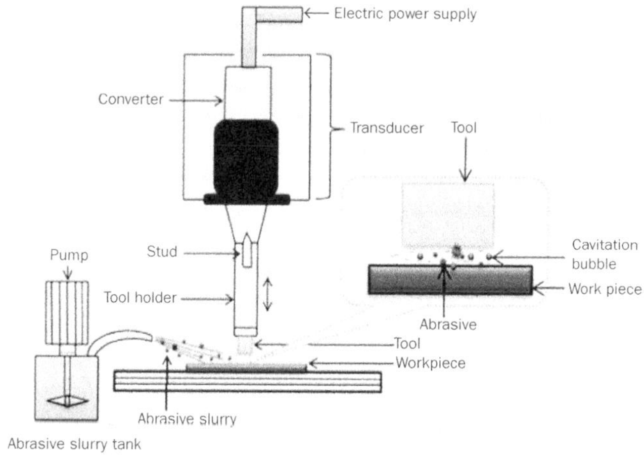

Figure 6.7 **Schematic diagram of ultrasonic set-up**

such as silicon carbide, diamond dust, aluminium oxide, boron carbide) acts as a cutting medium, takes away the wear particles and also cools the workpiece and tool.

As the abrasive slurry impacts the MMC surface, failure of the material occurs on the surface, and the liquid stream carries away both the abrasive particles and the cratered (wear out) particles. Most research has shown that the MRR in USM of MMCs is directly proportional to slurry pressure, abrasive concentration, tool vibration amplitude and grit size. An increase in MRR can be achieved with an increase in slurry pressure, abrasive concentration, tool vibration amplitude and grit size.

The most common applications of USM are in:

- cutting of industrial diamonds
- drilling holes in materials that are difficult to machine
- instrument making
- cutting holes that have curved profiles.

Limitations of USM include

- very low material removal rate/cutting rate
- extensive tool wear
- hole depth is presently limited to 2.5 times the diameter of the tool
- small depth of holes and cavities.

6.8.1.1 Input process parameters (USM) and output quality characterstics

Researchers have reported the effects of abrasive slurry, static force and amplitude of tool vibration on ultrasonic hole-making performance of MMCs. With an increase in the static force, MRR increases for various abrasive slurry, grain size and concentration. An increase in the amplitude of tool vibration causes an increase in dynamic force on the workpiece (Zhang *et al.*, 2000), but also leads to more effective flushing of debris (Thoe *et al.*, 1998). Increases in dynamic forces and effective flushing have a significant effect on MRR, and thus MRR increases with the amplitude of tool vibration. On the other hand, with an increase in static load and the amplitude of vibration, tool wear rate increases. The increased tool wear affects the resonance frequency of the USM, which causes a decrease in amplitude of vibration, thereby decreasing MRR. Therefore, the input parameter setting of static force and amplitude of vibration is a trade-off between MRR and tool wear rate.

An increase in the abrasive particle size causes an increase in MRR up to an optimal value of abrasive size. A further increase in abrasive size decreases MRR and may even cause a decrease in the surface quality of the workpiece. The increase in MRR with an increase in abrasive size results from the coarser abrasive particles causing more damage to the workpiece material during the striking process. However, with further increase in abrasive size, MRR decreases because the actual number of abrasive particles striking the workpiece decreases when the size of the abrasives is increased at the same level of slurry concentration (Hocheng *et al.*, 2000). An increase in abrasive size also causes an increase in tool wear rate and flushing of debris but a decrease in the surface quality of the workpiece.

6.8.2 Electrical discharge machining and drilling

EDM/EDD is a process based on thermal energy. It involves eroding the surface of a workpiece with a high-voltage spark. The prerequisite of EDM/EDD with respect to workpiece material is that it should be electrically conductive. An electrical potential difference is built up between the workpiece and the tool electrode. A dielectric fluid insulates the electrode from the workpiece until the voltage is sufficient to ionize the fluid and create a spark path through the fluid. At this point a high-temperature spark erodes a small crater from the surface of the workpiece. The current is then switched off and the eroded material is flushed away (Ghosh and Mallik, 2001). Schematic views of the EDM and EDD systems are shown in Figures 6.8 and 6.9, respectively. The working principle behind EDD is similar to that for EDM. The spark is generated between a rotary tool electrode and workpiece. EDD uses the high energy of the spark focused on the workpiece surface to remove material by localized melting and vaporization.

The main objective of EDD is to develop a hole-making technique to maximize MRR and minimize tool wear rate. The negative terminal is connected to the rotatory tool

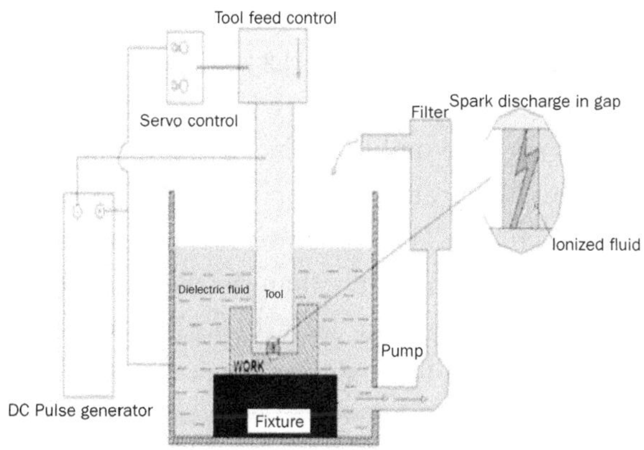

Figure 6.8 Basic scheme of electrical discharge machining

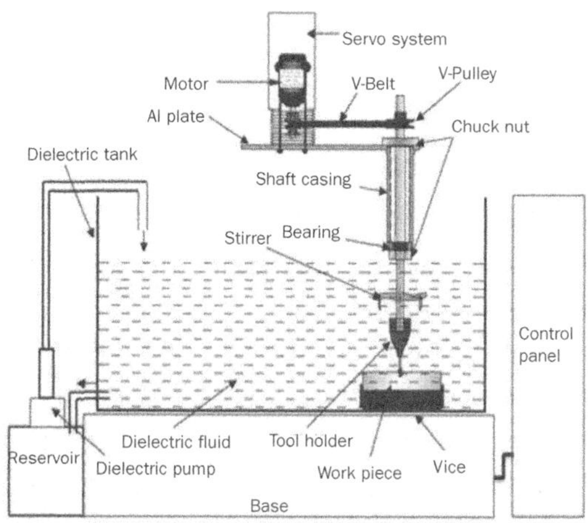

Figure 6.9 Basic scheme of electrical discharge drilling

(cathode) and positive terminal to the workpiece (anode). The tool and work surface are physically separated by the dielectric fluid. The gap between cathode and anode (10–125 μm) is controlled by a servo system to regulate the tool feed rate based on feedback signals of the average interelectrode gap voltage. Dielectric fluid is used as an effective cooling medium and it also removes the debris produced during hole making. The advantage of EDD lies in the continuous, accurate and precise hole-making operation at high productivity. The use of a rotating tool electrode is more productive and accurate than the use of a conventional static electrode. MRR and surface finish are higher for a rotary electrode than for a stationary one.

6.8.2.1 Input process parameters (EDM/EDD) and output quality characteristics

The mechanism and quality obtained with the EDM process are equivalent to those with the EDD process. The spark energy provided by the plasma channel melts and vaporizes the MMC workpiece (Yan and Wang, 1999). The factors affecting the output responses during the EDM of MMCs are (in descending order of importance) discharge current, pulse on-time, pulse off-time, gap voltage and flushing pressure (Dvivedi *et al.*, 2008).

MRR is higher for all rotating modes of the electrodes than for the stationary modes. This is possibly due to the superior debris removal effect of the rotating electrodes. Debris at the machining gap is removed effectively by the centrifugal force generated by the rotating solid electrode. However, the dielectric pressure generated is not sufficient to sweep the molten material, which directly reduces MRR during the EDM process. MRR is also dependent on the sparking energy, which is proportional to pulse duration (Yan *et al.*, 2000). Moreover, higher pulse on-time duration results in greater material removed per arc time, which results in a larger crater size (Singh *et al.*, 2004). Tool wear increases with an increase in the percentage of abrasive reinforcement. This increase in tool wear is due to the abrasive nature of the SiC particles. An increase in the percentage of SiC decreases conductivity, which increases tool wear rate. Discharge current has a direct effect on tool wear rate, suggesting that a very high discharge current can cause excessive tool wear.

The advantages of EDM/EDD include:

- no cutting force is generated between the tool and workpiece
- intricate details can be generated
- superior surface finish and accuracy
- smaller and deeper holes can be produced
- well suited for automation
- burr-free finish
- low-cost tooling
- materials that are difficult to machine, such as MMCs, can be machined.

Disadvantages include:

- low metal removal rate
- electrodes must be electrically conductive
- power consumption is high.

References

Basavarajappa S, Chandramohan G, Davim J P, Prabu M, Mukund K, Ashwin M, Prasanna Kumar M (2008), 'Drilling of hybrid aluminium matrix composites', *International Journal of Advanced Manufacturing Technology*, 35: 1244–50.

Basavarajappa S, Chandramohan G, Prabu M, Mukund K, Ashwin M (2007), 'Drilling of hybrid metal matrix composites – Workpiece surface integrity', *International Journal of Machine Tool and Manufacture*, **47**: 92–6.

Davim J P (2003), 'Study of drilling metal matrix composites based on Taguchi techniques', *Journal of Material Process Technology*, **132**: 250–4.

Dvivedi A, Kumar P, Singh I (2008) 'Experimental investigation and optimization in EDM of Al 6063 SiCp metal matrix composite', *International Journal Machining and Machinability of Materials*, 3(3/4): 293–308.

Ghosh A, Mallik A K (2001) *Manufacturing Science*. New Delhi: EWP Publications.

Hocheng H, Tai N H, Liu C S. (2000), 'Assessment of ultrasonic drilling of C/SiC composite material', *Composites: Part A Applied Science and Manufacturing*, **31**: 133–42.

Kilickap E (2010) 'Modeling and optimization of burr height in drilling of Al-7075 using taguchi method and response surface methodology', *International Journal of Advance Manufacturing Technology*, doi 10.1007/s00170-009-2469-x.

Monaghan J, Reilly P O (1992), 'The drilling of Al/SiC metal matrix composites', *Journal of Material Process and Technology*, **33**: 469–80.

Morin E, Masounave J, Laufer EE (1995), 'Effect of drill wear on cutting forces in the drilling of metal matrix composites', *Wear*, **184**: 11–16.

Mubaraki B, Bandyopdhyay S, Fowle R, Mathew P, Heath P J (1995), 'Drilling studies of an Al_2O_3 Al metal matrix composite. 1. Drill wear characteristics', *Journal of Material Science*, **30**: 6273–80.

Pandey P C, Shan H S (1980), *Modern Machining Processes*. Tata McGraw-Hill, pp. 7–38.

Ramulu M, Rao P N, Kao H (2002), 'Drilling of (Al_2O_3)p/6061 metal matrix composites', *Journal of Material Process Technology*, **124**: 244–54.

Riaz Ahamed A, Asokan P, Aravindan S, Prakash M K (2009), 'Drilling of hybrid Al-5% SiCp-5%B4Cp metal matrix composites', *International Journal of Advance Manufacturing Technology*, doi 10.1007/s00170-009-2453-5.

Singh P N, Raghukandan K, Rathinasabapathi M, Pai B C (2004), 'Electrical discharge machining of Al-10% SiCp as-cast metal matrix composites', *Journal of Materials Processing Technology*, **155–156**: 1653–7.

Thoe T B, Aspinwall D K, Wise M L H (1998), 'Review on ultrasonic machining', *International Journal of Machine Tools and Manufacturing*, **38**(4): 239–55.

Yan B H, Wang C C (1999), 'The machining characteristics of Al_2O_3/6061Al composite using rotary electro discharge machining with a tube electrode', *Journal of Materials Processing Technology*, **95**: 222–31.

Yan B H, Wang C C, Liu W D, Huang F Y (2000), 'Machining characteristics of Al_2O_3/6061Al composite using rotary EDM with a disk like electrode', *International Journal of Advanced Manufacturing of Technology*, **16**(5): 322–33.

Zhang Q H, Wu C L, Sun J L, Jia Z X (2000), 'Mechanism of material removal in ultrasonic drilling of engineering ceramics', *Proceedings of the Institution of Mechanical Engineers, Part B: Journal of Engineering Manufacture*, **214**(9): 805–10.

7

A laboratory machine for micro electrochemical machining

G. R. Ribeiro,[1] I. M. F. Bragança,[1] P. A. R. Rosa[1] and P. A. F. Martins[1]

Abstract: This chapter describes a laboratory machine for micro-electrochemical machining. The machine consists of a basic rigid structure, various electrical and electronic components, an electrolyte flow system and a numerically controlled tool positioning servomechanism. The machine was designed, fabricated and instrumented by the authors and, besides giving details for those readers who may be interested in developing low-cost equipment for fundamental research or teaching purposes, the chapter also emphasizes the adequacy of its operative conditions for drilling micro-holes of 500 μm in stainless steel AISI 304 sheets with 1 mm thickness under varying pulsed electric currents and concentrations of sulfuric acid.

Key words: Laboratory machine, micro-electrochemical machining, micro-holes, operative conditions.

7.1 Introduction

Electrochemical machining (ECM) is a non-traditional material removal process proposed by Gussef in 1929 that creates a

[1] Instituto Superior Técnico, Technical University of Lisbon, Av. Rovisco Pais 1049-001 Lisboa, Portugal.

Published by Woodhead Publishing Limited, 2013

contoured shape on a workpiece by means of controlled anodic dissolution of metals. In ECM the workpiece is immersed in an electrolyte and connected to the positive terminal of a low-voltage, high-current DC power supply while the tool electrode, a negative of the desired shape, is immersed at some distance (gap size) from the workpiece and connected to the negative terminal (Figure 7.1; Fan and Hourng, 2010).

ECM is the reverse of electroplating and was developed throughout the 1960s and 1970s into a production technology for the aerospace, automotive, biomedical and electrical industries, among others.

Although ECM may be considered a niche production technology it has several important advantages against traditional material removal processes, such as: (i) the longevity of tooling; (ii) the applicability to all conductive materials regardless of their strength, hardness, toughness and heat resistance; (iii) the ability to produce high-quality surfaces with close tolerances in workpieces with design features that are costly, difficult or even impossible to achieve by means of traditional machining processes; and (iv) the ability to ensure high production rates with

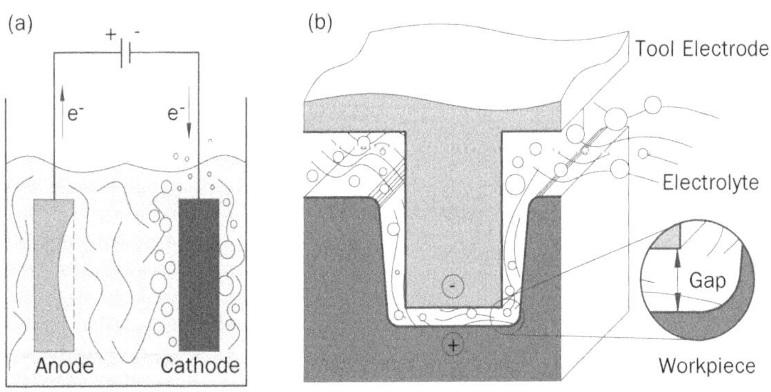

Figure 7.1 **Schematic representation of (a) the anodic dissolution of metals and (b) electrochemical machining with a contoured tool electrode**

repeatable results regardless of the operator (McGeough, 1974; Bhattacharyya *et al.*, 2004; Fan *et al.*, 2010).

Despite the aforementioned advantages, the widespread and effective utilization of ECM in industry has been considerably below expectation due to technical difficulties related to process stability, electrolyte and sludge removal, positioning and geometrical errors of the electrode, and stiffness of the machines (Bannard, 1977). All these problems affect the gap size between the electrode and the workpiece and strongly influence the overall accuracy and quality of the ECM products. These difficulties are particularly relevant when fabricating holes by ECM due to difficulties of the electrolyte in washing metal ions away from the gap before they have a chance to plate onto the electrode (Rajurkar *et al.*, 1999).

Besides the influence of process parameters related to the electrolyte, gap and current, ECM is also dependent on the reaction products of electrochemical machining and mass transport in the electrolyte. The electrolyte acts as a current carrier and must provide a high current efficiency η for the supplied current I to allow good material-removal rates (MRR):

$$\text{MRR} = C.I.\eta \qquad [7.1]$$

where C is a material constant and values of MRR typically range from 1 to 20 cm^3/h depending on current density J. Electrolytes must also guarantee sensitivity between current efficiency and the supplied current in order to ensure high accuracy and quality of the ECM components (Chikamori and Ito, 1971).

Mass transport in the electrolyte is governed by diffusion, migration and convection. Diffusion is the spontaneous movement of material particles from an area of high concentration to an area of low concentration, migration is the movement of charged material particles by the application of a voltage between the electrode and the workpiece, and convection is the movement of material particles in a stirred electrolyte.

The contribution of diffusion derives from the differences in concentration of ionic species in the electrolyte. The contribution of migration is proportional to the magnitude of the electric field between the electrode and workpiece, to the conductivity of the electrolyte, and to the concentration and charge of the ionic species. The contribution of convection to mass transport is caused by forced movement of the electrolyte and is described by hydrodynamics.

The flow rate and circulation of the electrolyte in the gap influences temperature (Joule heating) and the products of the electrochemical reaction (e.g. hydrogen and metallic ions). Joule heating increases MRR due to higher anodic dissolution kinetics and greater conductivity of the electrolyte. The release of hydrogen at the electrode decreases MRR because it dilutes the electrolyte concentration and diminishes its capacity to act as a current carrier. Metal ions from the reaction products also decrease MRR if they are not properly washed away from the workpiece surface (Bannard, 1977).

Tool making is among the manufacturing industries that have taken advantage of ECM for reducing or eliminating traditional finishing operations after hardening, namely the expensive hand-polishing of dies and moulds. The drilling of small holes is among the most cost-effective and successful applications of ECM in tool making due to the high quality and precision of the resulting surfaces.

The last few years have seen a growing demand for micro mechanical parts driven by a global trend toward product miniaturization. This movement has come not only from consumers, who want small and highly sophisticated electronic equipment, but also from recent applications in medicine, sensors and optoelectronics. Tool making for micro manufacturing applications opens news technological challenges by taking ECM a step forward to meet the precision requirements of complex geometric features in the micrometre range (Rajurkar *et al.*, 1999). The well-known and established ECM process and machine-tools cannot be simply scaled down to be applied on the micro scale. Scaling down requires identification of the key operating parameters of ECM at the micro-level together with innovative solutions and concepts for the design and fabrication of machines for micro-ECM (μECM).

This chapter presents the details of a laboratory machine for μECM that is capable of drilling holes with diameters down to 300 μm. The machine operates with ultra-low voltage, low electrolyte concentration, high-frequency short power pulses and a micro-tool electrode to localize the dissolution area and produce components with dimensions in the micrometre range. Experiments in drilling micro-holes validate the development and allow us to understand the influence of basic operative parameters such as current, frequency and electrolyte concentration in MRR and dimensional accuracy (DA) of the resulting holes.

7.2. Development of the laboratory machine

Figure 7.2 shows the laboratory machine for μECM as designed, fabricated and instrumented by us. The machine consists of four different groups of components: (i) basic structure, (ii) electrical and electronic parts, (iii) electrolyte flowing system and (iv) position control devices.

The structure was rigidly constructed by means of a metallic frame reinforced with marble plates and blocks to prevent vibrations and associated geometric errors.

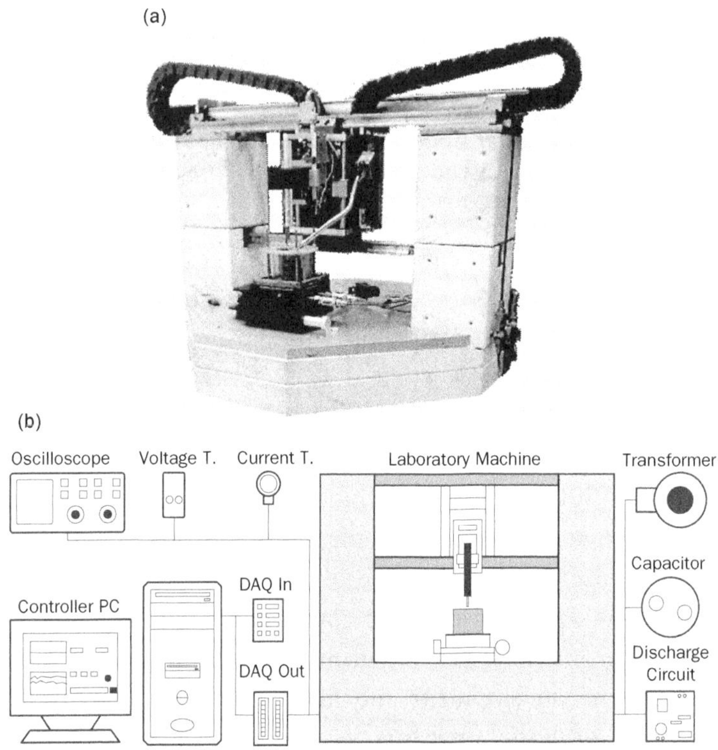

Figure 7.2 **The laboratory machine for μECM as designed and fabricated by the authors. (a) Photograph of the equipment and (b) schematic representation of the main components.**

The electrical and electronic parts comprise a DC power supply, a bank of capacitors and a discharge circuit. The DC power supply consists of a variable-voltage transformer and a constant-current rectifier for converting the single-phase 220-V AC to DC with voltage in the range 0–100 V and current rating up to 3 A. The bank of capacitors is charged by the DC power supply and its discharge is controlled by a metal-oxide semiconductor field effect transistor that generates square pulses of electric current with frequencies in the range 10 kHz to 1 MHz over a duty cycle (D).

$$D = \frac{t_{\text{on}}}{t_{\text{on}} + t_{\text{off}}} \quad [7.2]$$

equal to 0.5 (50 per cent). In the above equation, t_{on} is the pulse on-time and t_{off} is the pulse off-time in the period $T = t_{\text{on}} + t_{\text{off}}$.

The reason for including a transistor for generating pulses of electric current is because the precision of pulsed ECM is higher compared with direct-current ECM as a result of reaction products and Joule heat being efficiently removed from the gap during the pulse off-time (Datta and Landolt, 1981). The average current $\bar{I}$ in the period T is calculated from the amplitude of pulse current I as follows:

$$\bar{I} = \frac{t_{\text{on}}}{t_{\text{on}} + t_{\text{off}}} \cdot I \quad [7.3]$$

and is the quantity utilized in Eq. 7.1 whenever pulsed ECM is used.

The electrolyte flow system includes a pump, a filter and a tank. The pump forces the electrolyte to circulate through the gap between the electrode and workpiece in order to wash out the dissolved metal, to remove Joule heat and to ventilate the hydrogen that is formed in the electrode (cathode) during electrolysis. The tank is made from glass reinforced with stainless steel to prevent corrosion and the filter enables the electrolyte to be continuously reused by removing suspended solids.

Two position control systems are utilized. A manually operated *X*–*Y* table with digital micrometer heads and a *Z*-axis position servomechanism holds and feeds the electrode into the workpiece at a precise rate. The servo control of the *Z*-axis is performed by means of software developed in LabView that utilizes feedback signals of voltage and current obtained from a Hameg HZ100 voltage transducer and a

Bergoz CTB1.0 current transducer. The servo control is critical for keeping the gap constant and preventing the electrode from shorting out against the workpiece. The position servomechanism of the Z-axis is capable of moving the tool electrode along a total length of 120 mm with a precision of 0.5 μm.

A Tektronix 2004B oscilloscope is also available for observing and storing the nature of pulses and other related electrical phenomena that may occur during μECM.

7.3 Experimental work plan

Assessing the integration of the above components into a laboratory machine for μECM was performed by drilling micro-holes with diameters of 300 μm in stainless steel AISI 304 sheets with 1 mm thickness.

Electrolytic copper (DIN E-Cu58) wire with 0.3 mm diameter was utilized as electrode and the electrolyte was a low-concentration solution of sulphuric acid (H_2SO_4 at 96 per cent) that forms ions of hydrogen (H^+) and sulphate ions (SO_4^{2-}) as electrical current is applied between the electrode and workpiece. Metal ions react well with the sulphate ions in the electrolyte and the low concentration prevents corrosion of the tool electrode (Ahn, 2004).

The work plan was designed to isolate the influence of three main process parameters that were considered critical for analysing MRR and the dimensional accuracy of the micro-holes that were drilled with the machine (Table 7.1): (i) the applied current, (ii) the frequency of the square pulses of the electric current and (iii) the electrolyte concentration. By keeping the other parameters constant, i.e. (i) flow rate of electrolyte and (ii) shape of the electrode, it was possible to reduce the total number of parameters that influence the

Table 7.1 The plan of experiments

Amplitude of current (A)	1, 1.5 and 2
Frequency (kHz)	10, 100, 500 and 1000
Electrolyte concentration (mL/L)	0, 2.5, 5, 10, 30 and 60

process. The number of possible combinations of variables is otherwise quite large.

Experimental MRRs (mm³/h) were obtained from the ratio between the volume removed from the workpiece and the total drilling time (Figure 7.3). The dimensional accuracy (DA) of the micro-holes that were drilled by the machine was evaluated from the ratio between the difference of the entrance d_0 and exit d_1 diameters and the total length of the micro-hole (Figure 7.3):

$$DA = (d_0 - d_1)/h \quad [7.4]$$

The tapered contour of the micro-hole results in poor dimensional accuracy and is caused by an absence of lateral tool feed in order to compensate for growth of the side gap.

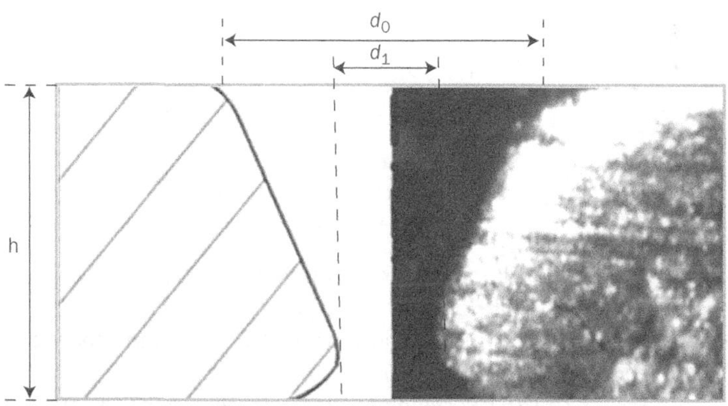

Figure 7.3 Tapered contour of the micro-holes utilized in the definition of dimensional accuracy (DA)

This phenomenon does not occur on the front gap due to the Z-axis position servomechanism and may be considered one of the major disadvantages of μECM.

7.4 Results and discussion

7.4.1 Electric current

Figure 7.4(a) shows MRR as a function of the amplitude of the pulses of electric current. The increase of MRR is not directly proportional to the current, as predicted by Faraday's

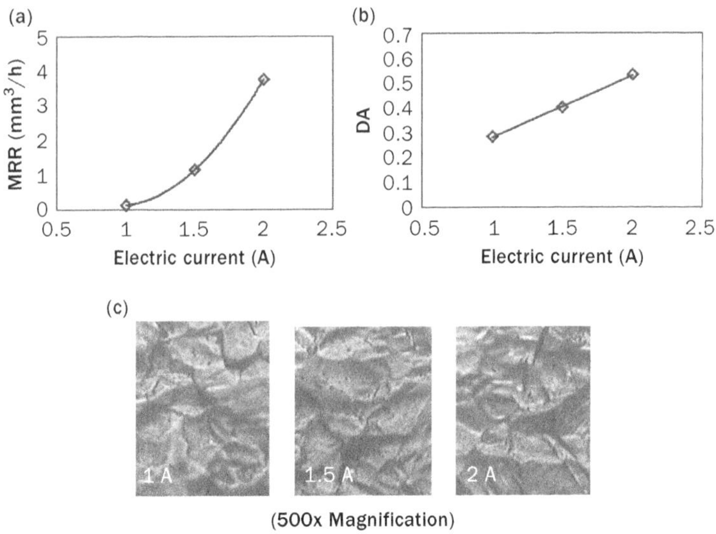

Figure 7.4 Influence of the amplitude of the pulses of electric current (frequency of 100 kHz) on (a) metal removal rate (MRR), (b) dimensional accuracy (DA) and (c) surface quality of the micro-holes. The electrolyte concentration is 60 mL/L

law of electrolysis (Eq. 7.1). Deviation from linearity increases as the current decreases due to the need to utilize very small gaps between the electrode and the workpiece and to the corresponding difficulties in removing Joule heat and washing metal ions away from the working region.

As a result, the leftmost region of Figure 7.4(a) is characterized by smaller values of current efficiency and current density, improving the overall dimensional accuracy of the micro-holes (Figure 7.4b). Regarding surface quality, there are no visible differences in the results obtained with various pulse currents (Figure 7.4c).

7.4.2 Frequency of the current pulses

In contrast to Figure 7.4(a), which shows a gradual decrease in MRR when the pulse current is reduced, the influence of frequency is much more sensitive (Figure 7.5a). Higher frequencies, leading to smaller pulse off-times t_{off}, rapidly decrease MMR. This is due to larger values of average current in the period T, which lead to greater amounts of heat and reaction products that can only be partially removed by the electrolyte during the pulse off-time, especially if the gap between the electrode and the workpiece is narrow.

The risk of metal ions being plated and contaminating the electrode increases with frequency and leads to non-uniform anodic dissolution and dimensional accuracy problems in the workpiece (Figure 7.5b). In extreme situations, the metal ions can even be trapped between the electrodes and stimulate micro-sparks that significantly decrease the overall surface quality of the micro-holes. As shown in Figure 7.5, the balance between MRR, DA and surface quality seems to be achieved at a frequency of 100 kHz.

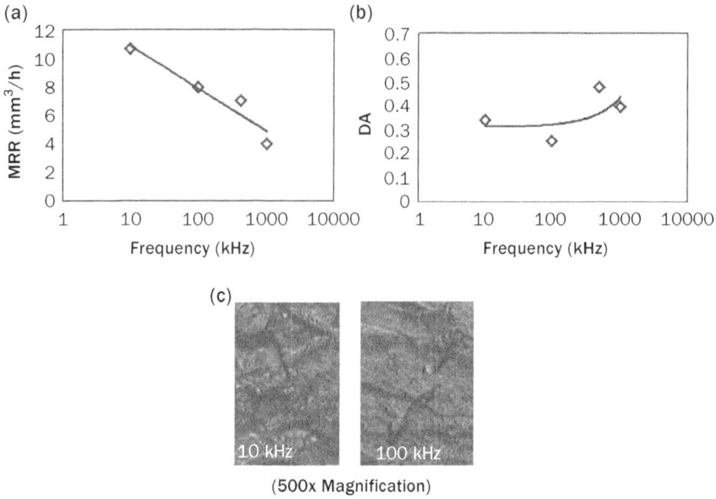

Figure 7.5 **Influence of the frequency of electric current pulses (amplitude of 2 A) on (a) metal removal rate (MRR), (b) dimensional accuracy (DA) and (c) surface quality of micro-holes. The electrolyte concentration is 30 mL/L**

The difficulty with stability of μECM when the frequency is raised from 100 to 1000 kHz is clearly seen in Figure 7.6 and indicates the necessity of diminishing the duty cycle or retracting the tool from the workpiece during the pulse off-time when using very high frequencies in order to enlarge the gap and intensify electrolyte flushing.

7.4.3 Electrolyte concentration

Figure 7.7(a) shows the evolution of MRR with the concentration of sulphuric acid (H_2SO_4) in the electrolytic solution. Two different regions are observed: a large region where MRR decreases as the electrolyte concentration increases and a very small region located at the leftmost area

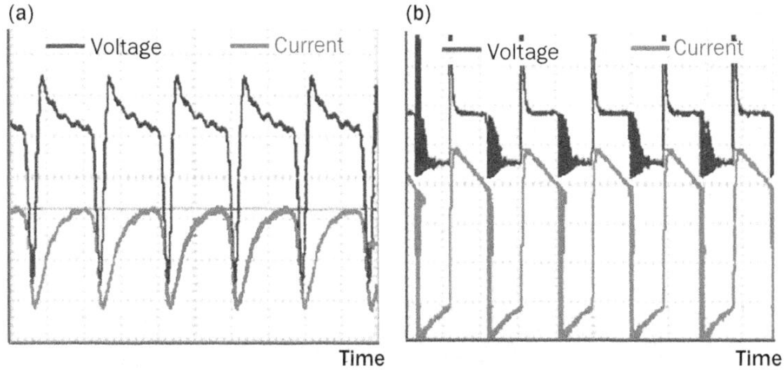

Figure 7.6 Acquired signals of current and voltage in the oscilloscope for pulses of electric current with an amplitude of 2 A and a frequency of (a) 1 MHz and (b) 100 kHz. The electrolyte concentration is 30 mL/L

of the graphic where MRR decreases sharply as electrolyte concentration decreases to zero.

The combined evolution of the large region in Figure 7.7(a) where MRR decreases as the electrolyte concentration increases with Figure 7.7(b) reveals that by properly choosing the electrolyte concentration it is possible to reach an optimal processing point (30 mL/L) that leads to the highest DA at the greatest possible MRR. The result is in close agreement with Kozak *et al.* (2000) and De Silva *et al.* (2003), who showed that by reducing the concentration away from the optimal processing point, the operating conditions will shift towards higher current densities and higher MRRs while increasing the concentration will do exactly the opposite.

However, the surface quality of the resulting micro-holes is better for lower electrolyte concentrations (Figure 7.7c).

In the leftmost area of the graphic, the results show that anodic dissolution is close to zero at very low electrolyte concentrations due to the absence of chemical action. In fact, when the electrolyte concentration is too low the only

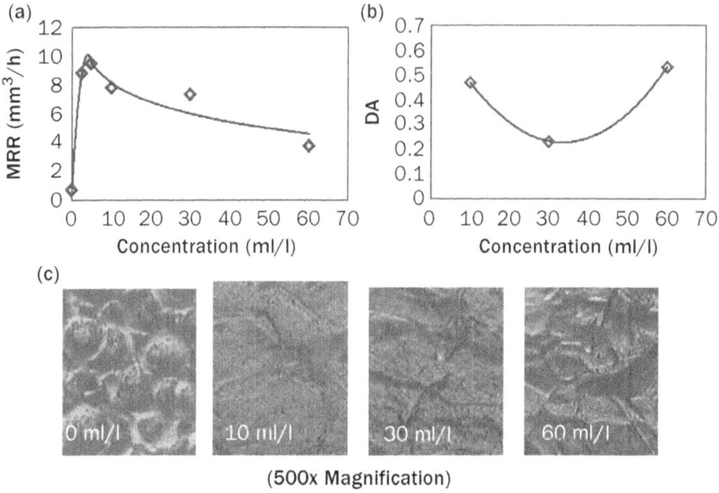

Figure 7.7 Influence of electrolyte concentration on (a) material removal rate (MRR), (b) dimensional accuracy (DA) and (c) surface quality of micro-holes for pulses of electric current (amplitude 2 A, frequency 100 kHz).

possibility of material being removed from the workpiece is by the intense heat generated by electrical discharges. As a result of this, surface quality decreases considerably (leftmost picture of Figure 7.7c).

Very low electrolyte concentrations require the physical characteristics and the material removal process to change from electrochemical to electrical discharge machining and water to be replaced by an appropriate dielectric fluid (electrically non-conducting fluid). The latter would act as an insulator while the voltage between the electrode and workpiece builds up sufficiently to generate a spark of intense heat (micro-plasma) through the gap.

From the above, we conclude that the proposed laboratory machine for μECM can also perform material removal by spark erosion and, therefore, can also be utilized for fundamental research and teaching in μEDM.

7.5 Conclusions

The proposed laboratory machine for μECM is capable of functioning across a broad range of operative conditions that are suitable for fundamental research and basic teaching purposes.

Variations in the amplitude and frequency of the pulses of electric current and in the concentration of sulphuric acid in the electrolytic solution give rise to significant differences in MRR, DA and quality of the micro-holes. However, it is possible to find an optimum processing point that leads to the highest DA at the greatest possible MRR for a specific electrolyte concentration.

The optimum processing point may be different from that ensuring the best surface quality, which was obtained for pulses of electric current with an amplitude of 2 A, a frequency of 100 kHz and an electrolyte concentration of 10 mL/L.

The proposed laboratory machine can be utilized for μEDM if the electrolytic solution is replaced by a dielectric fluid (e.g. mineral oil) and the operative parameters are properly adjusted. This is easily accomplished without modifying the structure, the electrical and electronic components, the liquid flowing system or the numerically controlled tool position servomechanism.

Acknowledgment

We would like to acknowledge Fundação para a Ciência e Tecnologia of Portugal for the financial support of this work under grants SFRH/BD/66576/2009 and SFRH/BD/48744/2008.

References

Ahn S (2004), 'Electro-chemical micro drilling using ultra short pulses', *Precision Engineering*, **28**: 129–34.

Bannard, J (1977), 'Electrochemical machining', *Journal of Applied Electrochemistry*, **7**: 1–29.

Bhattacharyya B, Munda J, Malapati M (2004), 'Advancement in electrochemical micro-machining', *International Journal of Machine Tools and Manufacture*, **44**: 1577–89.

Chikamori K, Ito S (1971), 'Electrolytic dissolution of mild steel in high rate current density region', *Denki Kagaku*, **39**: 493–8.

Datta M, Landolt D (1981), 'Electrochemical machining under pulsed current conditions', *Electrochimica Acta*, **26**: 899–907.

De Silva A K M, Altena H S J, McGeough J A (2003), 'Influence of electrolyte concentration on copying accuracy of precision-ECM', *CIRP Annals – Manufacturing Technology*, **52**: 165–8.

Fan Z-W, Hourng L-W (2010), 'Electrochemical micro-drilling of deep holes by rotational cathode tools', *International Journal of Advanced Manufacturing Technology*, **52**: 555–63.

Fan Z-W, Hourng L-W, Wang C-Y (2010), 'Fabrication of tungsten microelectrodes using pulsed electrochemical machining', *Precision Engineering*, **34**: 489–96.

Kozak J, Dabrowski L, Lubkowski K, Rozenek M, Slawinski R (2000), 'CAE–ECM system for electrochemical technology of parts and tools', *Journal of Materials Processing Technology*, **107**: 293–9.

McGeough J A (1974), *Principles of electrochemical machining*. London: Chapman and Hall.

Rajurkar K P, Zhu D, McGeough J A, Kozak J, De Silva A (1999), 'New developments in electrochemical machining', *CIRP Annals – Manufacturing Technology*, **48**: 567–79.

Published by Woodhead Publishing Limited, 2013

8

Cam-driven electromagnetic mechanical testing machine

C. M. A Silva,[1] P. A. R. Rosa[1] and P. A. F. Martins[1]

Abstract: This chapter describes an innovative testing machine for the mechanical characterization of materials under high rates of loading. The machine consists of an electromagnetic actuator, a fixed housing containing two flat compression platens, a translating cam and a follower. The electromagnetic actuator enables high strain rates to be reached with very precise control of the impact velocity and of the energy transmitted to the translating cam. The cam profile enables compression testing to be performed under the strain rate vs. strain loading paths that are commonly found in metalworking in order to meet the combined specifications of the machine-tool and process. Emphasis is placed on giving constructive details for those readers who are interested in developing a low-cost machine for conducting material tests at high rates of loading, as well as on demonstrating the flexibility and adequacy of its operative conditions by determining the flow curves of aluminium AA1050-O and technically pure lead (99.9 per cent) under different testing conditions. These are implemented in a finite-element computer program and applied in the numerical simulation of two different basic metalworking operations.

Key words: Compression testing, electromagnetic, flow curve, high strain rate.

[1] Instituto Superior Tecnico, Technical University of Lisbon, Av. Rovisco Pais 1049-001 Lisboa, Portugal.

Published by Woodhead Publishing Limited, 2013

8.1 Introduction

The flow curve at the appropriate rates of loading is important in describing the hardening behaviour during plastic deformation in terms of strain, strain-rate and temperature, to set-up the non-linear constitutive equations of metal plasticity and to establish the feasibility window and effectiveness of metalworking processes. However, despite this importance, the flow curve is not always accessible in conditions similar to real metalworking due to difficulties in replicating the operative conditions, namely the combined evolution and range of strains and strain-rates.

In the case of strain-rates, for example, the widely available universal testing machines can only perform mechanical characterization of materials at quasi-static or low rate loading conditions ($\dot{\bar{\varepsilon}} \sim 10^{-3}$ to $10\ s^{-1}$) and commercially available drop weight testing systems are only adequate for medium rates of loading ($\dot{\bar{\varepsilon}} \sim 1$ to $10^3\ s^{-1}$). Special purpose testing equipment based on split Hopkinson pressure bars, which is adequate for high rates of loading ($\dot{\bar{\varepsilon}} \sim 10^3$ to $10^4\ s^{-1}$), and Taylor impact systems, which are necessary for even higher rates of loading ($\dot{\bar{\varepsilon}} > 10^4\ s^{-1}$), are not readily available to purchase or guaranteed to the majority of researchers.

However, even if equipment for testing at high rates of loading is available, a frequently ignored and mismanaged technical issue is the adequacy of testing conditions to the operational settings of the machine-tools where metalworking processes will take place. In fact, despite theoretical claims on transferability of results often requiring mechanical testing of materials with different strain and strain-rate loading profiles to provide the same stress response for a given value of strain and strain rate, the flow curves obtained under test with unclear conditions lead to non-unique and non-representative responses for the same material (Banabic *et al.*, 2000). The time history of strain and strain-rate influences the stress response (Silva *et al.*, 2010), and it is relevant to determine crystallographic textures resulting from metalworking processes (Guo *et al.*, 2011).

As a result of this, mechanical characterization of materials requires deep insight into the machine-tools and deformation mechanics of the processes prior to choosing the most appropriate equipment and operative test conditions (Field *et al.*, 2004). Taking metal cutting as an example, the synergism between the mechanical behaviour of materials and the machine-tools requires a good understanding of the displacement–time relationship of the cutting tool and of its influence in the rate-dependent variables of the process. Because strain-rate increases with the level of strain as material moves from the undeformed region to the shear plane (a very narrow plastic deformation zone around AB in Figure 8.1a) and decreases while material moves away from the shear plane up the rake face of the tool, the resulting strain-rate vs. strain loading path for a typical flow route (E in Figure 8.1a) is plotted in Figure 8.1(d).

This means that the typical strain-rate vs. strain loading paths of metal cutting are significantly different from those obtained with commercially available testing equipment for medium and high rates of loading. In the case of split Hopkinson pressure bars, for example, strain-rate vs. strain loading paths are characterized by an approximately constant level of strain-rate (Silva *et al.*, 2012), similar to that plotted for blanking in Figure 8.1(b,d), while for drop-height testing systems the strain-rate vs. strain loading paths are similar to that of forging (Figure 8.1c,d).

The above-mentioned difficulties in obtaining the flow curves at appropriate rates of loading and the aforesaid synergism between material testing and real metalworking

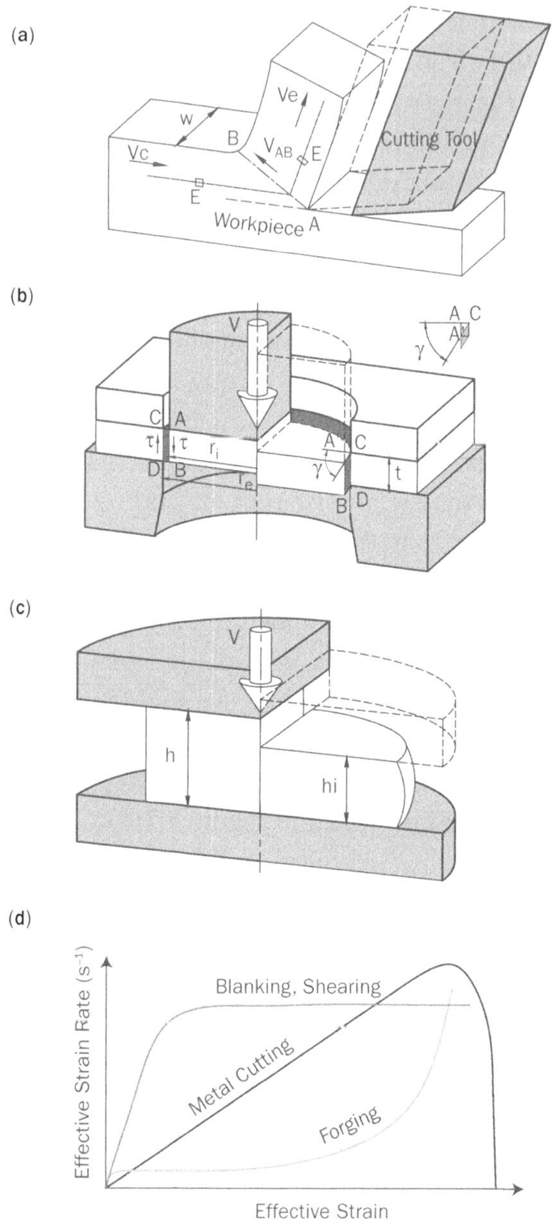

Figure 8.1 **Schematic representation of: (a) metal cutting, (b) blanking and shearing, (c) forging and (d) strain-rate vs. strain loading paths**

conditions justify our two-fold objective here: (i) the development of an innovative cam-driven electromagnetic machine for the compression testing of materials under high rates of loading and (ii) the identification of new testing methodologies based on the selection of the strain-rate vs. strain loading paths that can easily and effectively replicate the strain-rate vs. strain loading paths found in real metalworking.

The chapter ends with examples of finite element modelling in which the operative conditions of the cam-driven electromagnetic mechanical testing machine are shown to be flexible and adequate to replicate the deformation mechanics of metal cutting and shearing.

8.2 Design and fabrication of the testing machine

Figure 8.2 shows the cam-driven electromagnetic testing machine designed, fabricated and instrumented by us. Three main groups of components can be identified: (i) basic structural parts, (ii) specific mechanical parts and (iii) specific electrical and electromagnetic parts.

8.2.1 Basic structural parts

The basic structural parts comprise the support frame and fixing accessories that are independent of the type of testing conditions and materials to be characterized. The support frame allows installation of the specific mechanical, electrical and electromagnetic parts. The overall length of the frame was chosen so that larger compression housings and longer actuators can be added in the future with the objective of

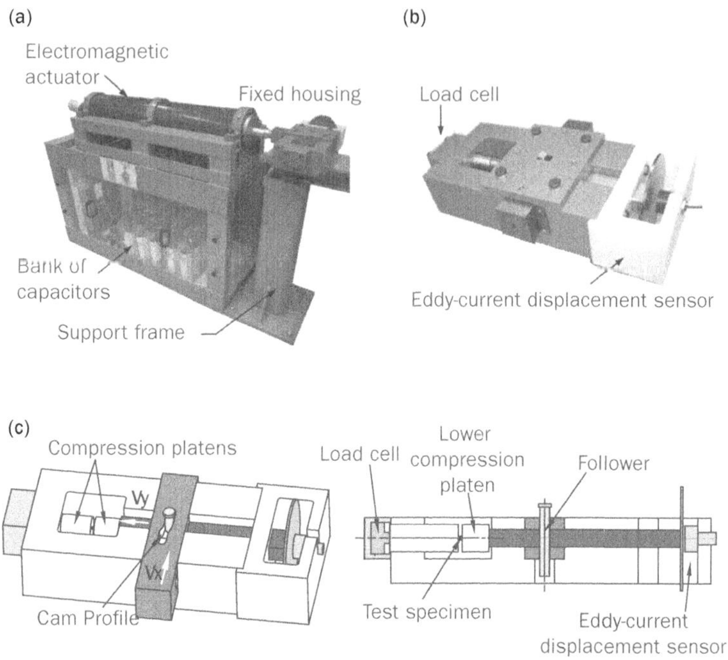

Figure 8.2 Cam-driven electromagnetic mechanical testing machine. (a) Photograph of the testing machine; (b) detail of the fixed housing; (c) schematic drawing showing the translating cam, follower, compression platens, load cell and displacement sensor

enlarging the size of the test specimens and increasing the velocity and kinetic energy of the ram.

8.2.2 Specific mechanical parts

Specific mechanical parts comprise a fixed housing containing two flat compression platens, a translating cam and a follower whose design depends on the operative conditions of the metalworking process to be replicated.

The compression platens are made from a cold working tool steel DIN 120WV4 hardened and tempered to 60 HRc. The translating cam and the follower are made from steel

DIN 14NiCr14 and DIN 100Cr6, respectively. The clearance fit for the cam–follower system is H7/f7 (ISO) and the individual parts were manufactured in a CNC machining centre. Final manual grinding and polishing was necessary to eliminate small surface errors and imperfections that although being imperceptible to the eye could cause high stress and vibrations in the cam follower. Protection of the sliding elements of the cam system was performed by means of a polytetrafluoroethylene (Teflon added) oil-based lubricant.

The surface contour of the cam (cam profile) is designed with the objective of synchronizing the performance of the testing machine with that of the machine-tool where the metalworking process will take place. The follower traces the cam profile and converts horizontal movement (x) of the ram to vertical displacement (y) of the lower compression platen (Figure 8.2c). The conversion of movement is schematically illustrated in Figure 8.3(a).

In the case of the cam shown in Figure 8.3(b), the profile (hereafter 'logistic profile') is characterized by an entry dwell followed by a rise contour and a final dwell towards the uppermost profile of the cam. The vertical displacement y of the follower as a function of the horizontal displacement x of the ram is depicted in Figure 8.3(c).

The velocity v_y of the follower is directly related to the first derivative of the displacement curve because the velocity of the ram v_x is approximately constant in the working region of the cam follower ($v_x^{\text{avg}} = 10$ m/s; between the dashed vertical lines in Figure 8.3d):

$$v_y = \frac{\mathrm{d}y}{\mathrm{d}t} = v_x \frac{\mathrm{d}y}{\mathrm{d}x} \cong v_x^{\text{avg}} \frac{\mathrm{d}y}{\mathrm{d}x} \qquad [8.1]$$

The leftmost region in Figure 8.3(d), characterized by a sharp increase in the velocity of the ram, results from the initial acceleration due to the pressure generated by the coils inside the electromagnetic actuator. There is no vertical

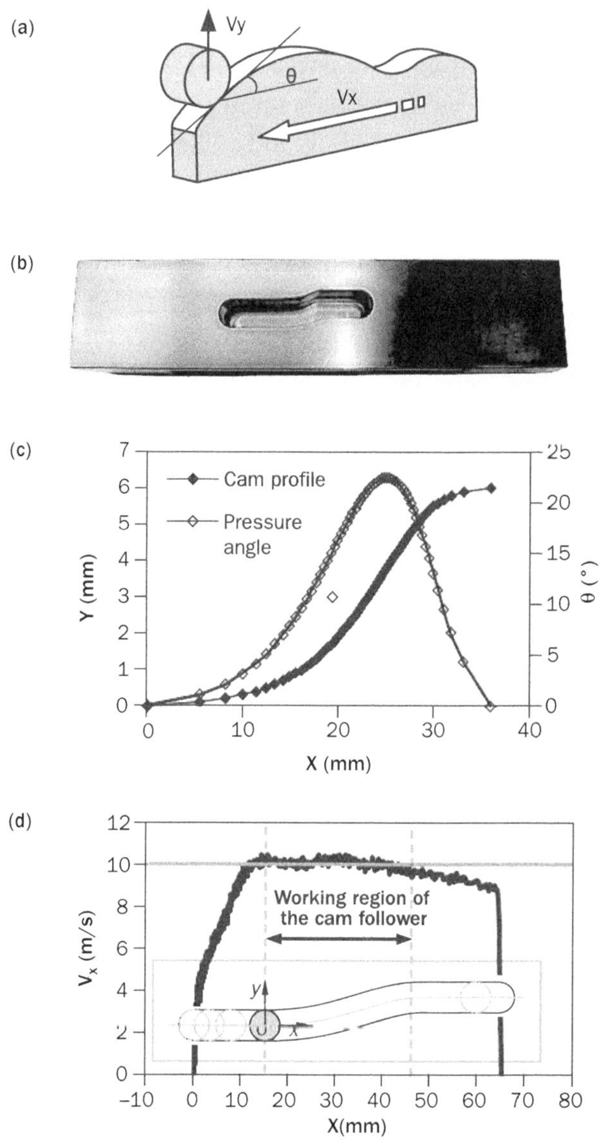

Figure 8.3 Testing machine equipped with a logistic cam. (a) Schematic representation of the cam profile and follower; (b) photograph of the logistic cam; (c) cam profile and pressure angle; (d) velocity of the ram v_x^{avg} = 10 m/s in the working region of the cam follower

movement of the cam follower along this region and therefore there is no compression of the upset test specimen. The rightmost region in Figure 8.3(d) shows part of the deceleration of the ram after the follower has reached the uppermost profile of the cam. Again, there is no compression of the upset test specimen along this region.

As a result of this, the vertical movement of the cam follower leading to upset compression takes place only at the region between the dashed vertical lines in Figure 8.3(d), hereafter termed the 'working region of the cam profile'. Thus, subsequent figures showing the displacement of the ram as the horizontal axis (x-axis) are frequently limited to the working region of the cam profile.

The acceleration a_y of the cam follower is computed from the variation in the velocity v_y (assuming the above mentioned approximation of v_x) and the kinematic analysis of the follower shown in Figure 8.4(a) allows us to conclude that the cam used in the experiments provides the maximum velocity at the point of inflection located near the maximum slope of the cam profile.

Acceleration is approximately constant at the entry dwell of the cam profile and presents an abrupt change from positive and negative values at the midpoint. Along the entry dwell of the cam, jerk is approximately null (Figure 8.4b).

The low acceleration at the entry dwell of the logistic cam profile, combined with the fact that pressure angles θ of the follower are kept below 30° (Rothbart, 2004) (θ_{max} = 22.5°, Figure 8.3c), helps to keep inertia forces at a low level and explains why the proposed testing equipment worked smoothly without shocks and vibrations while performing material testing at high rates of loading.

The logistic cam profile allows the material flow conditions to be replicated under conditions similar to those found in metal cutting processes (Silva *et al.*, 2012). However, because

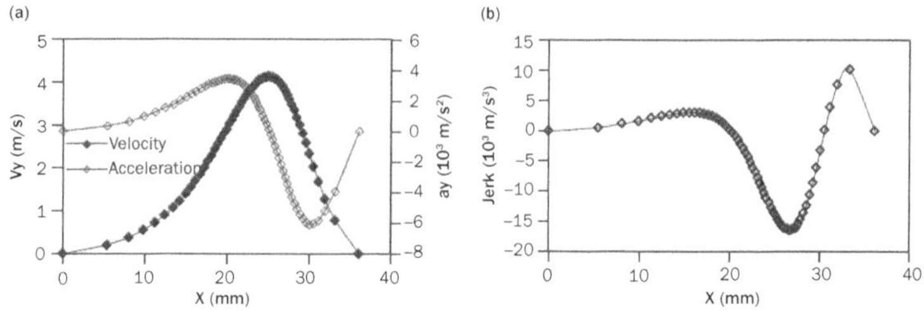

Figure 8.4 Kinematics of the testing machine equipped with a logistic cam. (a) Velocity and acceleration of the follower; (b) jerk in the working region

the cam system is flexible and its profile depends on the operative conditions to be replicated, it is easy to change the kinematics of the proposed equipment to replicate another metalworking process, machine-tool or material testing equipment. For instance, replacing the logistic cam by a 'root type cam' allows the machine to replicate the kinematics of a split Hopkinson pressure bar (Figure 8.5).

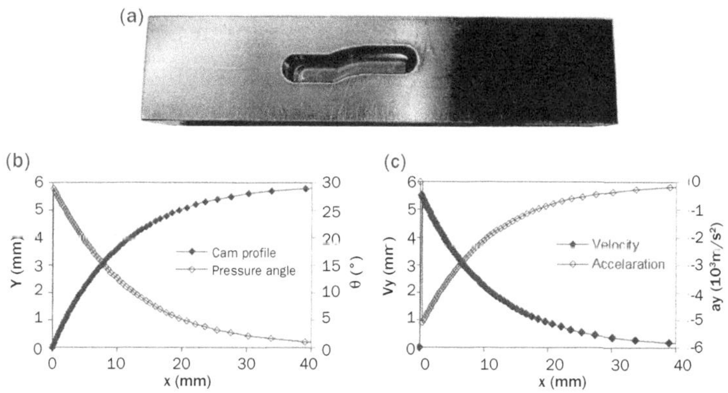

Figure 8.5 Testing machine equipped with a root-type cam. (a) Photograph of the root-type cam; (b) cam profile and pressure angle; (c) velocity and acceleration of the follower in the working region

Published by Woodhead Publishing Limited, 2013

8.2.3 Specific electric and electromagnetic parts

Specific electrical and electromagnetic parts include the eddy-current displacement sensor, the force transducer and the components that provide the energy to the electromagnetic actuator, e.g. the electrical circuits for charging and firing the bank of capacitors and the coils that generate the pressure for accelerating the ram linked to one end of the translating cam.

The vertical movement of the cam follower is transformed into compression of the test specimen via the lower flat platen. The vertical displacement of the lower flat platen is measured by an eddy-current displacement sensor and the resulting force is measured by means of a load cell connected to the upper compression platen (Figure 8.2c).

The electromagnetic actuator consists of electrical circuits for charging and firing the banks of energy-storage capacitors (each with 6 mF) and a series of coils that generate the pressure to accelerate the ram linked to the translating cam. Typical coils with eight windings, a total length of 92 mm, an external diameter of 160 mm and an internal diameter of 70 mm were utilized. The ram consists of a long (1.5 m) and heavy (5.2 kg) bimetallic bar made of AA 6082-T651 aluminium with slotted hollow half-ring surface inserts (to reduce eddy current losses) of DIN St52.3 steel (Figure 8.6).

The capacitors are charged by means of a single-phase 230-V AC that is converted to a higher voltage DC by means of a charging circuit consisting of a variable-voltage transformer (capable of producing 3.6 times the input voltage) and a constant-current rectifier system. Once the capacitors are charged, the charging circuits are closed and the thyristor switches located in the discharging circuits are activated to simultaneously fire each capacitor into its

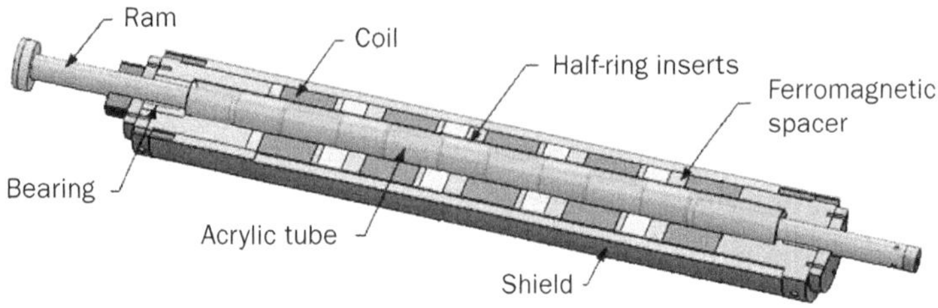

Figure 8.6 Electromagnetic actuator showing the ram, coils, half-ring inserts and ferromagnetic spacers

associated coil. The resulting current pulse will only last for a few milliseconds, not long enough to accelerate the ram to a velocity of 18 m/s.

8.3 Experimental workplan

The stress–strain curves of aluminium AA1050-O and technically pure lead (99.9 per cent) were obtained by means of compression tests on cylindrical specimens of 6 mm

Table 8.1 The plan of experiments for aluminium AA1050-O

Case	Test conditions	V_x	Case	Test conditions	V_x
1	Quasi-static	0.01	7	Root type cam profile	3.5
2	Logistic cam profile	2	8	Root type cam profile	7
3	Logistic cam profile	3.9	9	Root type cam profile	10.4
4	Logistic cam profile	5.8	10	Root type cam profile	14
5	Logistic cam profile	7.8	11	Root type cam profile	17.5
6	Logistic cam profile	9.2			

Table 8.2 The plan of experiments for technically pure lead (99.9%)

Case	Test conditions	V_x	Case	Test conditions	V_x
1	Quasi-static	0.01	7	Root type cam profile	3.7
2	Logistic cam profile	2	8	Root type cam profile	7.2
3	Logistic cam profile	3.8	9	Root type cam profile	10.9
4	Logistic cam profile	5.5	10	Root type cam profile	14.5
5	Logistic cam profile	7.2	11	Root type cam profile	17.6
6	Logistic cam profile	8.5			

diameter and 6 mm height. Both logistic and root-type cam profiles were utilized and the quasi-static conditions were included as a reference. Details of the experimental workplan are presented in Tables 8.1 and 8.2.

8.4 Results and discussion

This section starts by presenting the mechanical characterization of aluminium AA1050-O and technically pure lead under high rates of loading and ends by evaluating the performance of the logistic and root-type cam profiles in modelling the non-proportional loading paths that are commonly found in orthogonal metal cutting and shearing.

8.4.1 Mechanical characterization

Figure 8.7 shows the variation of force with displacement obtained from test cases 3 and 8 in Table 8.1. In contrast to root-type cam profiles (and split Hopkinson pressure bars),

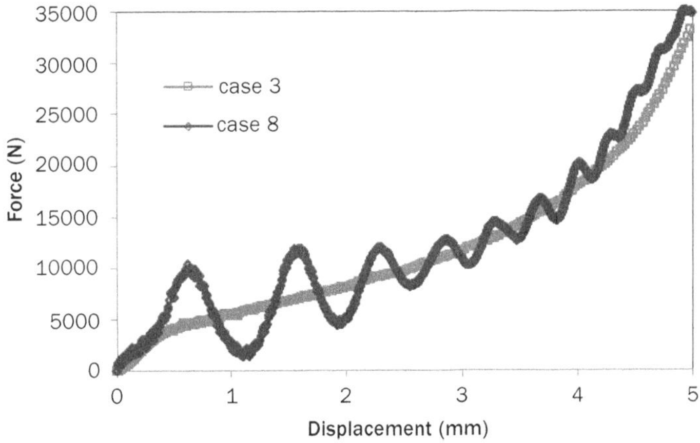

Figure 8.7 Experimental evolution of the force vs. displacement for logistic and root-type cam profiles in the mechanical testing of aluminium AA1050-O

there are no 'saw tooth' oscillations when using a logistic cam profile (case 3). This is attributed to a smooth transition between the entry dwell and the profile of the cam and to a low maximum pressure angle ($\theta^{max} = 22.5°$, Figure 8.3c).

The oscillations in the root-type cam profile are attributed to inertia forces and stress wave propagation under impact loading as well as to an initial value of the maximum pressure angle very close to 30° ($\theta^{max} = 28°$, Figure 8.5a).

The experimental strain-rate vs. strain loading paths for the selected test conditions performed with the cam-driven electromagnetic testing machine equipped with a logistic cam profile are plotted in Figure 8.8(a).

The three-dimensional flow surface plotted in Figure 8.8(b) in which stress, strain and strain-rate are the leading axis results from fitting experimental data to the following mathematical material model developed by us (Silva *et al.*, 2012):

$$\sigma = \left(A + e^{m}\,\varepsilon^{n}\right)\left(B + C\ln\left[D + \varepsilon\right]\right) \qquad [8.2]$$

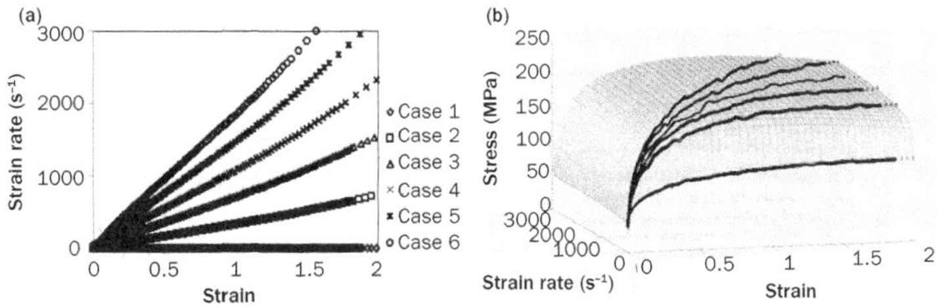

Figure 8.8 **Mechanical testing of aluminium AA1050-O using a logistic cam. (a) Strain-rate vs. strain loading paths; (b) material stress response with respect to strain and strain-rate (experimental data and fitting)**

where the constants A, B, C, D, m and n are determined from the experimental data.

The model given by Eq. 8.2 is appropriate for cold forming operating conditions and includes well-known material models such as those of Ludwik–Holloman, Voce and Johnson–Cook (isothermal), among others, as special cases. Another advantage of this model is the ability to exhibit material flow softening at high values of strain. Flow softening is responsible for diminishing the resistance to plastic deformation due to rearrangement of dislocations under dynamic recrystallization and is known to have a significant influence on the deformation mechanics of metal cutting, namely chip formation (Sima and Özel, 2010).

Figure 8.9(a) provides similar results for the root-type cam. The strain-rate vs. strain loading paths obtained with this type of cam are near horizontal lines and similar to those commonly attained with Hopkinson pressure bars. The three-dimensional flow surface resulting from these experiments is plotted in Figure 8.9(b).

The differences between the flow surfaces obtained from material testing with logistic and root-type cams are made clear by analysing a selection of intersections of the

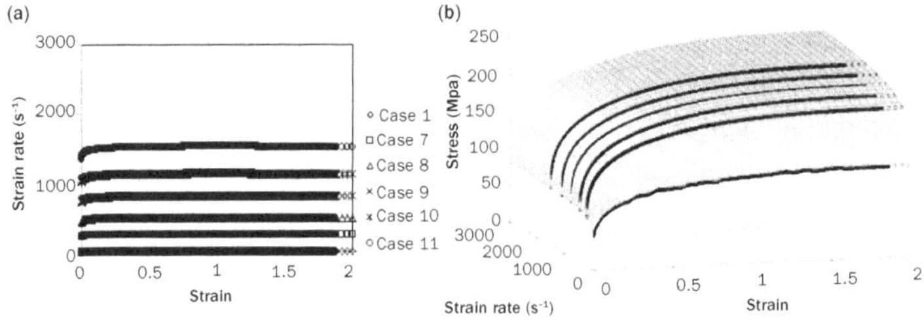

Figure 8.9 Mechanical testing of aluminium AA1050-O using a root-type cam. (a) Strain-rate vs. strain loading paths; (b) material stress response with respect to strain and strain-rate (experimental data and fitting)

three-dimensional flow surfaces with constant strain-rate planes ($\dot{\varepsilon} = Cte$). The flow curves resulting from these intersections are plotted in Figure 8.10 and indicate that major differences are due to flow softening, in close agreement with the aforementioned ability of the logistic cam profile to model metal cutting conditions.

The flow surfaces for technically pure lead obtained from material testing with logistic and root-type cams are shown

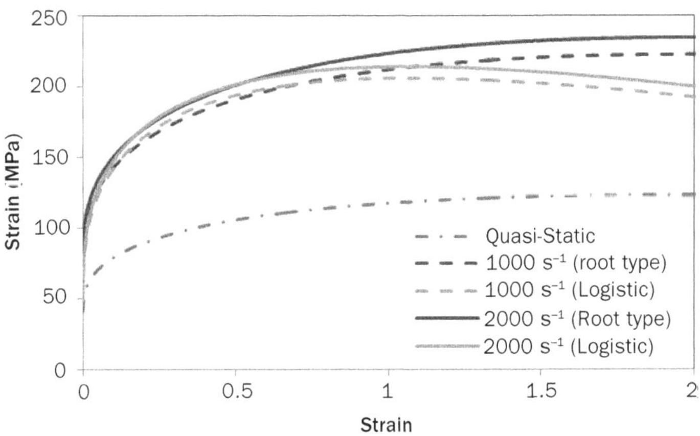

Figure 8.10 Experimental stress–strain curves for aluminium AA1050-O for different values of strain-rate

Published by Woodhead Publishing Limited, 2013

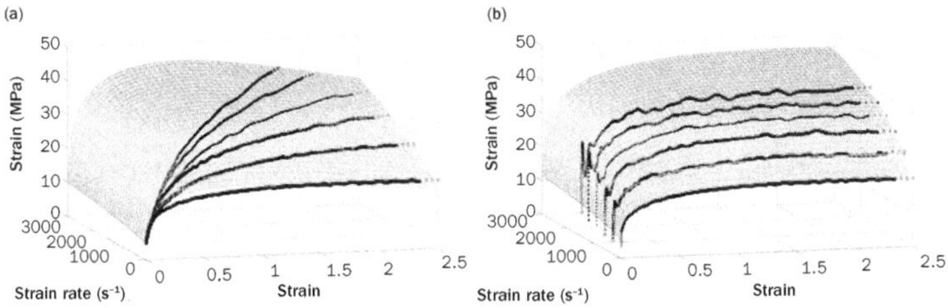

Figure 8.11 **Material stress response of technically pure lead with respect to strain and strain-rate (experimental data and fitting) using (a) a logistic cam and (b) a root type cam**

in Figure 8.11. The procedure followed that utilized for aluminium AA1050-O.

8.4.2 Orthogonal metal cutting and shearing

Figure 8.12 shows the initial and deformed finite-element model after reaching steady-state conditions in orthogonal metal cutting. The model was designed to vary the cutting speed and maintain the uncut chip thickness (0.25 mm) and the clearance angle of 5° and rake angle of 10° of the cutting tool throughout the simulations. The cutting specimen comprised a parallelepiped of technically pure lead of 10 × 5 × 20 mm.

As seen in Figure 8.12(b) the computed strain-rate vs. strain paths for different flow routes (see detail in Figure 8.12a) in the plastically deforming region match those of the proposed equipment operating with a logistic cam. Conversely, the loading paths derived from finite-element analysis reveal the inadequacy of mechanical characterization of materials for metal cutting applications being performed with root-type cams (which, in general

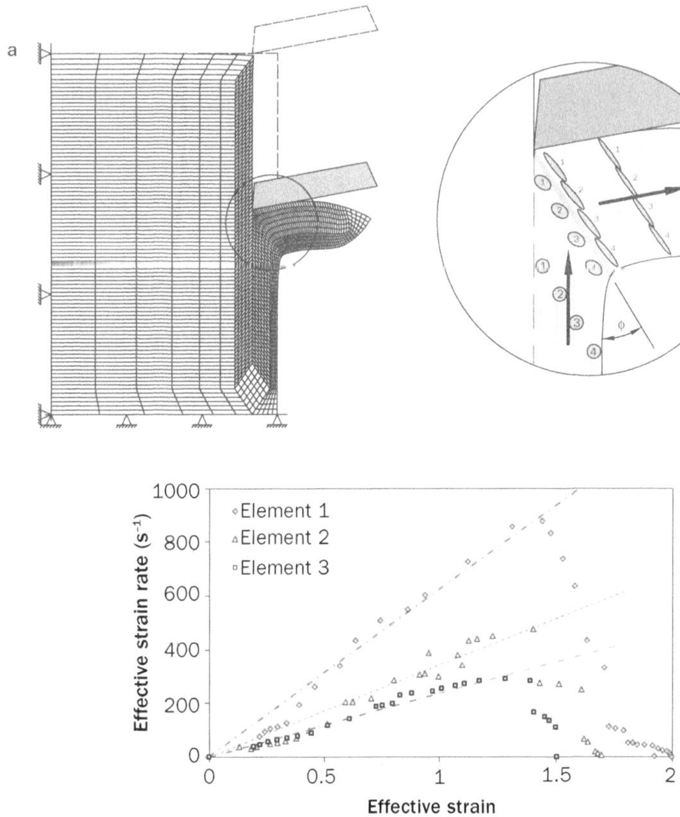

Figure 8.12 **Finite-element modelling of orthogonal metal cutting. (a) Mesh and detail of the plastically deforming region; (b) computed strain-rate vs. strain loading paths for different flow routes along the plastically deforming region**

terms, behave as conventional split Hopkinson pressure bars).

Figure 8.13 shows the initial and deformed finite-element model at the onset of cracking in a double-notched cylindrical test specimen loaded under shear. The fracture test specimen consisted of a cylinder of aluminium AA1050-O of 12 mm height and 30 mm initial diameter. As seen in Figure 8.13(b),

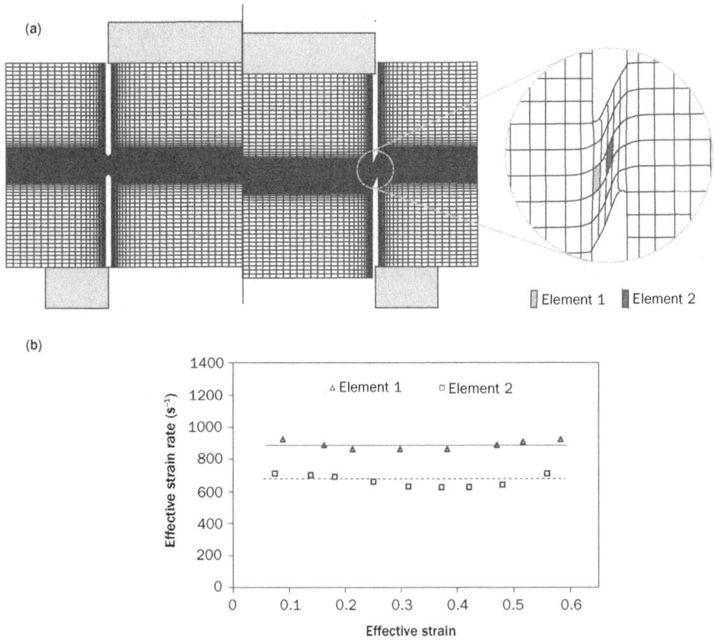

Figure 8.13 **Finite-element modelling of double-notched cylindrical test specimens loaded in shear. (a) Mesh and detail of the plastically deforming region; (b) computed strain-rate vs. strain loading paths for two different elements located in plastically deforming region**

and in contrast to orthogonal metal cutting, the root-type cam is more appropriate to model the non-proportional loading paths of shearing.

8.5 Conclusions

The proposed cam-driven electromagnetic testing machine is capable of operating across a broad range of operative conditions found in real metalworking processes. Variations in cam profiles, modifications in the individual coils of the actuator, which can be easily changed or temporarily switched off during testing, and changes in the operating voltage of the charging and discharging circuits (which influence the kinetic energy of the ram) can be easily and effectively performed to achieve different strain vs. strain-rate loading paths.

Experiments show that the testing machine equipped with a logistic cam is capable of diminishing 'saw tooth' oscillations in the experimental recordings of force that are typical of split Hopkinson pressure bars and root-type cam profiles. Moreover, logistic cam profiles in conjunction with the proposed material model are able to successfully model the flow softening that is commonly found in metal cutting applications. On the other hand, root-type cam profiles are shown to be adequate to model plastic flow under shearing.

Acknowledgment

We would like to acknowledge Fundação para a Ciência e Tecnologia of Portugal for the financial support of this work under research grant SFRH/BD/46798/2008.

References

Banabic D, Bunge H-J, Pöhlandt K, Tekkaya A E (2000), *Formability of Metallic Materials*. Berlin: Springer-Verlag.

Field J E, Walley S M, Proud W G, Goldrein H T, Siviour C R (2004), 'Review of experimental techniques for high rate deformation and shock studies', *International Journal of Impact Engineering*, 30: 725–75.

Guo Y, Saldana C, Mann J B, Saoubi R M, Chandrasek S (2011), 'Deformation and microstructure in machining', *Advanced Materials Research*, 223: 325–31.

Rothbart H A (2004), *Cam Design Handbook*. New York: McGraw Hill.

Silva C M A, Rosa P A R, Martins P A F (2010), 'Mechanical characterization of materials for bulk forming using a drop weight testing machine', *Journal of Mechanical Engineering Science*, **224**: 1795–804.

Silva C M A, Rosa P A R, Martins P A F (2012), 'Electromagnetic cam driven compression testing equipment', *Experimental Mechanics*, **52**: 1211–22.

Sima M, Özel T (2010), 'Modified material constitutive models for serrated chip formation simulations and experimental validation in machining of titanium alloy Ti-6Al-4V', *International Journal of Machine Tools and Manufacture*, **50**: 943–60.

Published by Woodhead Publishing Limited, 2013

Index

acceleration, 219–22
acknowledgements
acoustic emission (AE)
  amplitude, 11–16
  FFT. *See* AE-FFT
  RAW, 8, 10, 13–14
  RMS, 3–5, 8, 13–15, 25
  sensors, 3–4, 8
  signals, 1–25
  STFT method, 19–25
AdvantEdge™ software, 49, 52, 60
AE-FFT
  of a cutting pass with four divided subsets, 10–11
  from different cutting passes, 9, 16–17
  intensity, 9–11, 16
  spectra, 2, 6, 8–11, 13–15, 25
  of subsets of a cutting pass, 17–19
AISI303 steel, 37
  chemical composition, 43–4, 73–4
  effect of cutting speed in high-speed machining, 62–86
  mechanical properties, 43–4
  micro-hardness of, 63, 71
  photomicrographs of, 45, 63
  validation tests, 49–62
AISI304 steel, 34–5
aluminium AA1050-O
  fracture test for, 228
  mechanical characterization of, 223–7
  plan of experiments for, 222
  stress–strain curves for, 226
amplitude, 11–16, 18–24, 67, 204, 206–9
analysis of variance (ANOVA), 138–9
  for MRR, 152
  for surface roughness, 153
asymmetric rotors, 95–7
austenitic stainless steels, 29–87
  EDX of, 72–4, 77, 83
  machinability of, 36–40
  machining study of, 43–86
  properties of, 33–43
  SEM of, 68–9, 72–3, 75–6, 78–80, 82
  turning, 32–3, 48–62
axial forces, 100–1, 105

Box–Behnken design, 147–51
built-up edges (BUEs), 38–9, 178

calibration
  curves, 101
  of torque, 99–101
  for workpiece holder, 99–101
cam-driven electromagnetic mechanical testing machine, 211–30
  design and fabrication of, 215–22
  experimental workplan, 222–3
  with logistic cam, 217–20, 223–7, 230
  photograph of, 216, 218, 220
  with root type cam, 220, 223–30
cam follower, 217–18, 220–1
cam profile, 217–20, 223–30
chip formation, 79–86, 225
chip thickness, 58, 69–70, 83, 85–87, 106–8, 227
  non-deformed, 59–61, 86
chip–tool interface, 41, 43, 49, 53, 55–6
compressed air jet, 126–7
compression, 219
  platens, 216
  testing, 215, 222–3
compressors, 92
  screw rotors, 95–7, 109
conventional cutting fluids, 114
  advantage of, 118–20
  grinding in, 119
conventional hole making, 169, 171, 174–82, 184–5
conventional machining methods, 137, 143–4
cooling-lubrication, 115–16, 125, 127
corrosion
  inter-grain, 35
  resistance to, 32–4, 36, 38, 68
cubic boron nitride (CBN), 93–5
  grinding, 94, 127
  wheels, 113, 121–2
current pulses, 204–6, 209
cutting conditions, 32–3, 53, 60–2
cutting edge, 69–70, 73–4, 77, 79, 178, 180–2
cutting fluid
  composition, 116
  conventional, 114, 118–20
    See also conventional cutting fluids
  reduction, 113–14
  usage, 115–18, 127–9
cutting forces, 41–2, 45–6, 53–4, 59–65, 87, 128, 179–80
cutting inserts, 6–9, 21–4
cutting passes, 9–25
cutting speed, 29–30, 32–3, 37–43, 49, 53–87, 107, 179–82, 227
  chip thickness, 59, 70
  in high-speed machining, 62–86
  influence on
  chip formation, 79–86
  cutting forces, 63–5
  part surface integrity, 68–72
  superficial finish, 65–8
  tool wear, 72–9
  tangential cutting force and, 63–4

cutting time, 9–19, 22–5, 178
cutting tool, 43, 72–4, 128, 181

delamination, 1–3, 13, 21–2, 24–5
depth of cut, 106–8, 119, 126
diamond-coated cutting tools, 1–25
die-sinking EDM, 140–3, 145
dielectric fluid, 135–6, 141–2, 146, 189–90, 208–9
dimensional accuracy (DA), 199, 202–9
double-notched cylindrical test, 228–9
drilling, 169, 171
  electro-discharge, 183–4, 188–91
  holes, 198–9
  micro-holes, 199, 202–5
  of MMCs, 176–82

electric current, 202, 204–5
electrical discharge machining (EDM), 135–64, 169, 171, 182–4, 188–91
  applications of, 145
  characteristics, 138–9
  die-sinking, 140–3, 145
  electrodes, 139
  input process parameters, 190–1
  micro, 141, 143–4
  output quality characterstics, 190–1
  parameters, 138–9, 147–52, 156–64
  principles of, 140–5
  wire-cut, 141–4
electro-chemical machining (ECM), 182–3, 195–9
  direct-current, 201
  micro, 195–209
  pulsed, 201
electro-discharge drilling (EDD), 183–4, 188–91
  input process parameters, 190–1
  output quality characterstics, 190–1
electrodes, 135–6, 139–43, 146, 190, 197–199, 201–2, 205
  rotation, 148, 152, 158–64
electrolyte, 197–8, 201–2, 204–5
electromagnetic actuator, 221–2
energy-dispersive X-ray spectroscopy (EDX), 72–4, 77, 83
experimental tests
  set-up and methods, 6–12
  work plan, 202–4, 222–3
  for workpiece, 101–3

fabrication
  of cam-driven electromagnetic mechanical testing machine, 215–22
  processes, 172–4
fast Fourier transformation (FFT) analysis, 2, 6, 8–11, 13–15, 25
feed rate, 60–2, 65, 106–8, 178, 180–2
finite element analysis, 48–9, 60, 227–9

flow curves, 211–13, 226
flow softening, 225–6, 230
flushing pressure, 148, 152, 159–64
force
  components
    idle strokes without material removal, 104
    trends of, 104–8
  grinding, 94–5, 101–3, 109, 125
  monitoring, 91–109
fracture test, 228–9

graphite, 139, 179–82
grinding, 114, 116–18
  CBN wheels, 113, 121–2, 127
  MQL, 119, 123, 129
  parameters, 103, 107–8
  strategies, 96–7
  wheels, 93–4, 102–3, 124

high-performance machining, 29–87
high-speed machining (HSM), 30, 40–3
  effect of cutting speed in, 62–86
high-speed steel (HSS), 177–8, 181
hole making
  conventional, 169, 171, 174–82, 184–5
  in MMCs, 176–91
  unconventional, 169, 171, 174–5, 182–91

input parameters, 51–2, 187–8, 190–1

laboratory machine, 195–209
logistic cam, 217–20, 223–7, 230

machining, 1–25, 114. *See also* electrical discharge machining; electro-chemical machining
  of austenitic stainless steels, 29–87
  cam-driven electromagnetic mechanical testing, 211–30
  conditions, 7, 13–14, 32–3
  conventional methods, 137, 143–4
  experiments, 145–9
  non-conventional methods, 136–7
  non-traditional, 135–6
  parameters, 138–9, 147–52, 156–64
  problems with cutting fluid usage, 116–18
  tests, 12–24, 49–63, 72–3
  ultrasonic, 182–3, 185–8
  WC/Co composites, 135–64
material removal rate (MRR), 138–40, 146–9, 152–64, 197–9, 202–9
matrix material, 171–3
mechanical processes, of unconventional machining, 182–5
mechanical testing
  of aluminium AA1050-O, 224–6
  of technically pure lead, 227
Merchant theory, 85–6

metal cutting, 214–15, 219, 223, 225–6, 230
metal matrix composites (MMCs), 169–91
  classification of, 171–3
  drilling, 176–82
  fabrication processes for, 172–4
  hole making in, 176–91
  primary processing, 172–3
  secondary processing, 172–3, 175–6
    development of, 170–1
  ultrasonic machining, 169, 182–3, 185–8
micro-electrochemical machining (μECM), 195–209
minimum quantity lubrication (MQL), 113–15

non-conventional machining methods, 136–7

output quality characteristics, 187–8, 190–1

peak current, 148, 152, 156, 158–64
photograph
  of logistic cam, 218
  of root type cam, 220
  of testing machine, 216
plastic deformation, 57, 80, 176, 225
polycrystalline diamond (PCD), 2, 139, 177–9
pressure angles, 217–18, 220, 224
primary processing, 172–3
pulse off-time, 205–6
pulse on-time, 148, 152, 158–64

radial forces, 99–101
residual analysis, 154–6
residual graphs
  for MRR, 154
  for surface roughness, 155
response surface methodology (RSM), 139–40
  for MRR and surface roughness, 152–6
  procedure for, 151
  using Box–Behnken design, 149–51
root-mean-squared (RMS), 3–5, 8, 13–15, 25
root type cam, 220, 223–30

scanning electron microscopy (SEM), 68–9, 72–3, 75–6, 78–80, 82
screw rotors
  generality of, 95–7
  grinding, 95–7
  manufacturing of, 92
secondary cutting fluid jet, 115, 125–6
shape grinding, 91–109
shearing, 214–15, 223, 227–9
  angle, 53–4, 85–6
short-time Fourier transform (STFT) method, 2, 6, 11–13, 19–25
simulation tests, 49, 51–3, 60

split Hopkinson pressure bars, 220, 223, 228, 230
stainless steels
  austenitic. *See* austenitic stainless steels
  characterization, 31–2, 80
  machining of, 36–40
  production, 31
strain, 211–13, 224–7
strain rate, 56–7, 64–5, 211–15, 224–7
stress–strain curves, 222, 226
surface quality, micro-holes of, 205–9
surface roughness, 65–8, 128, 140, 146–56, 180–2
  of ANOVA, 153
  effect of, 160–3
  residual graphs, 155
symmetric rotors, 95–7

tangential cutting force, 53, 63–4, 126
technically pure lead
  mechanical characterization of, 223, 226–7
  plan of experiments for, 222–3
thermal damage, 119, 122
  grinding in, 120–1
thermal processes, of unconventional machining, 182–5
tool geometry, 39–40, 59, 180–2
tool wear, 43, 47–8, 70, 72–9, 128, 139, 176–8, 190–1
torque
  calibration, 99–100
  grinding parameters for, 108
  idle strokes without material removal, 104
  trends of, 104–8
tungsten carbide (WC), 139–40, 177–8
tungsten carbide/cobalt (WC/Co) composites, 135–64
turning, 32–3, 48–62

ultrasonic machining (USM), 169, 182–3, 185–8
unconventional hole making, 169, 171, 174–5, 182–91

wheel cleaning, 113, 115, 122–8
wheel clogging, 113, 115, 122–4
workpiece, 122–4, 140–3, 146–7, 198, 201–2, 205–6, 208
  holder, 97–101
  micrograph of, 160–1
  mounting, 101–3
  rotation, 104–8